AF768649

# Iron Ores and Iron Oxides – New Perspectives

*Edited by Brajesh Kumar*

Published in London, United Kingdom

Iron Ores and Iron Oxides - New Perspectives
http://dx.doi.org/10.5772/intechopen.103994
Edited by Brajesh Kumar

Contributors
Shaista Ali, Aliya Zahid, Syeda Taskeen Shahid, K. Lakshmanan Palanisamy, Kesavan Vignesh, Nanjappan Karthikeyan, Nancy Jaswal, Purnima Justa, Hemant Kumar, Deepshikha, Krishna, Balaram Pani, Pramod Kumar, Zuoliang Zhang, Dayanand Paswan, Malathi Madhurai, Joseph Ekhebume Ogbezode, Olusegun Olufemi Ajide, Oluleke Olugbemiga Oluwole, Olusoji Ofi

© The Editor(s) and the Author(s) 2023

The rights of the editor(s) and the author(s) have been asserted in accordance with the Copyright, Designs and Patents Act 1988. All rights to the book as a whole are reserved by INTECHOPEN LIMITED. The book as a whole (compilation) cannot be reproduced, distributed or used for commercial or non-commercial purposes without INTECHOPEN LIMITED's written permission. Enquiries concerning the use of the book should be directed to INTECHOPEN LIMITED rights and permissions department (permissions@intechopen.com).

Violations are liable to prosecution under the governing Copyright Law.

Individual chapters of this publication are distributed under the terms of the Creative Commons Attribution 3.0 Unported License which permits commercial use, distribution and reproduction of the individual chapters, provided the original author(s) and source publication are appropriately acknowledged. If so indicated, certain images may not be included under the Creative Commons license. In such cases users will need to obtain permission from the license holder to reproduce the material. More details and guidelines concerning content reuse and adaptation can be found at http://www.intechopen.com/copyright-policy.html.

Notice

Statements and opinions expressed in the chapters are these of the individual contributors and not necessarily those of the editors or publisher. No responsibility is accepted for the accuracy of information contained in the published chapters. The publisher assumes no responsibility for any damage or injury to persons or property arising out of the use of any materials, instructions, methods or ideas contained in the book.

First published in London, United Kingdom, 2023 by IntechOpen
IntechOpen is the global imprint of INTECHOPEN LIMITED, registered in England and Wales, registration number: 11086078, 5 Princes Gate Court, London, SW7 2QJ, United Kingdom

British Library Cataloguing-in-Publication Data
A catalogue record for this book is available from the British Library

Additional hard and PDF copies can be obtained from orders@intechopen.com

Iron Ores and Iron Oxides - New Perspectives
Edited by Brajesh Kumar
p. cm.
Print ISBN 978-1-83769-544-7
Online ISBN 978-1-83769-543-0
eBook (PDF) ISBN 978-1-83769-545-4

# We are IntechOpen, the world's leading publisher of Open Access books

## Built by scientists, for scientists

**6,500+**
Open access books available

**177,000+**
International authors and editors

**195M+**
Downloads

**156**
Countries delivered to

Our authors are among the
**Top 1%**
most cited scientists

**12.2%**
Contributors from top 500 universities

WEB OF SCIENCE™

Selection of our books indexed in the Book Citation Index in Web of Science™ Core Collection (BKCI)

Interested in publishing with us?
Contact book.department@intechopen.com

Numbers displayed above are based on latest data collected.
For more information visit www.intechopen.com

# Meet the editor

Dr. Brajesh Kumar is an assistant professor and head of the Department of Chemistry, TATA College, Kolhan University, Chaibasa, India. He received a Ph.D. in Chemistry from the University of Delhi, India, in 2009. He is a pioneering researcher in green chemistry, nanosynthesis, environmental remediation, photocatalysis, nanomedicine, polymers, and natural product extraction. Dr. Kumar has several national and international fellowships to his credit and has worked as a faculty member at various universities throughout India, Ecuador, and South Korea. He is a holder of two registered patents as well as a member of the American Chemical Society and the Indian Society of Chemists and Biologists (ISCB). He has also published more than eighty indexed research articles and is an active reviewer of more than seventy journals. He was listed among the world's top 2% of scientists by Stanford University, USA, in 2021 and 2022.

# Contents

# Preface

Iron oxide has numerous applications in food science, sensors, adsorbents, catalysis, data storage, electronic devices, the environment, engineering, drug-delivery technology, biomedical research, magnetic recording devices, and more. Throughout the last few decades, extraction of iron oxide from mines using various techniques has been investigated by many researchers. Nanoscience and nanotechnology will continue to grow due to the numerous benefits they confer in our daily lives, including in human health, environmental safety, and device engineering. This will continue to encourage scientists to develop iron or iron oxide nanoparticles whose nanocomposites have revealed many exciting and promising applications in environmental science, engineering, nanomedicine, catalysis, sensing, therapeutic agents, food science, and beyond.

This book discusses the extraction, synthesis, characterization, and applications of iron, iron oxide, and their nanoparticles/nanocomposites. It examines existing methods as well as novel green synthetic methods that have been developed to circumvent some of the problems of the traditional synthesis of iron oxide nanoparticles. It also highlights characterization methods, including current trends in spectroscopic, microscopic, and diffraction methods.

Written by experts, chapters address fundamental assumptions involved in describing the reactions, structure, application, and toxicity of iron ores and iron oxide derivatives.

In Chapter 1, Shaista Ali et al. focus on the engineering of iron nanoparticles using several environmentally friendly techniques and their potential to remove contaminants from the environment. They emphasize the use of environmentally friendly agents such as polymers, amino acids, bacteria, fungi, plant extracts, and so on for the production of iron nanoparticles and correlate their particle size, morphology, and other aspects with their various applications.

In Chapter 2, K. Lakshmanan Palanisamy et al. highlight the application of nano-dimensional iron oxide nanoparticles with nanotoxicology. Iron oxide nanoparticles such as hematite ($\alpha$-$Fe_2O_3$) and maghemite ($\gamma$-$Fe_2O_3$) have been widely applied in various applications due to their particle size and surface area.

In Chapter 3, Nancy Jaswal et al. discuss biomedical applications of superparamagnetic iron oxide nanoparticles (SPIONS), including their use as theranostic agents. Magnetic nanoparticles like SPION are considered to be the most viable and frequently used in biomedical applications due to their low cost and lack of toxicity. Within the developing field of nanomedicine, SPIONs are basic technological classes that have been widely studied for cancer imaging and treatment.

Among the three iron oxides of hematite, magnetite, and wüstite, the latter is the most difficult to reduce. Pure wüstite is not found in nature because it is not stable in thermodynamics.

In Chapter 4, Zuoliang Zhang examines the reduction thermodynamics and kinetics of wüstite by C, CO, and $H_2$ in a blast furnace. When the temperature is lower than 810°C (1083 K), the reduction ability of CO is stronger than that of $H_2$. If it is higher than 810°C (1083 K), the opposite is true. The use of pure hydrogen in iron making in the future is desirable.

In Chapter 5, Dayanand Paswan and Malathi Madhurai investigate the feasibility of the process for reducing ilmenite ore pellets using lean-grade non-coking coal and the separation of $TiO_2$ from iron via melting at the laboratory scale. Conventionally, $TiO_2$ is produced from ilmenite ore by carbothermic reduction followed by leaching of iron to produce high-purity $TiO_2$, which leads to loss of metallic iron and disposal of effluents.

Finally, in Chapter 6, Joseph Ekhebume Ogbezode et al. provide a systematic overview of the different iron-making routes, their operational limitations, and future perspectives to mitigate the challenges of iron production demands in modern-day metallurgical processes. Thus, in today's iron and steel-making processes, non-contact direct reduction and reduction-smelting processes remain viable alternative routes to iron production.

Finally, I am grateful to my beloved wife Kumari Smita for her helpful comments on several chapters and excellent support in conceptualizing the book's content.

**Dr. Brajesh Kumar**
Department of Chemistry,
TATA College,
Kolhan University,
Chaibasa, Jharkhand, India

Chapter 1

# Green Engineering of Iron and Iron Oxides by Different Plant Extract

*Shaista Ali, Aliya Zahid and Syeda Taskeen Shahid*

## Abstract

Green synthesis has emerged as a promising and eco-friendly approach for the synthesis of nanoparticles, including iron and iron oxide nanoparticles. This method utilizes plant extracts, microorganisms, or other natural resources as reducing and stabilizing agents instead of toxic chemicals to produce nanoparticles with desired properties. One advantage of green synthesis is the ability to control the size, shape, and crystalline structure of the nanoparticles, which can be analyzed using techniques such as powder X-ray diffraction (XRD). The two variable oxidation states (+2 and + 3) provide an opportunity for multiple products with different designing and crystallite structures. Iron and Iron oxide nanoparticles are both important for biological and photochemical activities. The method for green synthesis decides what kind of particles will one get and for what activity it is suitable. Iron nanoparticles are more suitable for biological activities like antibacterial, antimicrobial, anti-cancerous, and iron oxide for photoelectrical like band gap studies, conduction and photo-catalysis.

**Keywords:** sustainable green nanotechnology, zerovalentiron-nanoparticles, plant extract, pXRD, zeta potential

## 1. Introduction

Nanochemistry is a branch of chemistry that focuses on the synthesis, characterization, and manipulation of nanoparticles and nanomaterials. Over the past few years, the field of nanoscience and technology has gained increasing importance in research and development areas. Nanoparticles and nanomaterials have diverse properties that can be tuned to meet specific requirements, making them useful in various applications, such as electronics, energy storage devices, and drug delivery systems. Nanoparticles can also be used for environmental remediation, water treatment, and catalysis, among others [1].

Iron is hard and brittle metal, as the fourth most abundant element in the earth's crust; it makes up the majority of the planet's core by weight. Researchers believe that iron has been used by humanity for about 5000 years. The most commonly found sources of iron are the minerals magnetite and hematite. However, smaller quantities of iron can also be derived from the minerals siderite, taconite, and limonite. It contains four fundamental structures, alpha, beta, gamma, and delta, where particles join under various circumstances [2].

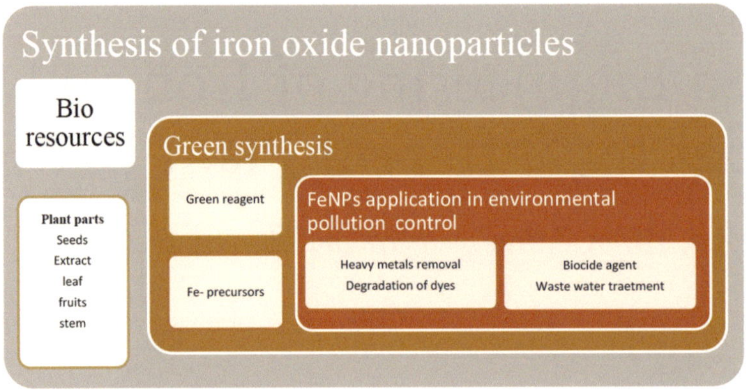


**Figure 1.**
*Graphical abstract of green synthesis.*

**Ferrites** are categorized into four groups: alpha (attractive), beta, gamma, and omega. The symbol for iron in the periodic table is Fe, which has a nuclear weight of 55.845 and a nuclear number of 26 with a thickness of 7.874 g/cm$^3$. Iron has high melting and boiling point of 1538 and 2862°C, respectively, making it a useful material for applications where high temperatures are involved, such as in the manufacturing of steel or in metallurgical processes. There are 33 different isotopes in existence, but only four of them are stable. The most well-known is iron-56, which has a usual overflow of 91.75% [2].

Iron has been the subject of several recent scientific efforts due to its lithophilicity and chalcophilicity, which result in the production of minerals such as pyrite ($FeS_2$), magnetite ($Fe_3O_4$), hematite ($Fe_2O_3$), siderite ($FeCO_3$), and others. Ferritin, a type of protein that stores iron in the body, also contains iron oxide ($Fe_3O_4$). Magnetite, in particular, has garnered significant attention in recent years. It is not uncommon to find magnetite in the Magnesian area of Asia Minor, where it occurs in two forms: reduced ferrous iron species and oxidized ferric iron. This intriguing chemical has the potential to be employed in a range of applications and exhibits a variety of unique properties. The features of magnetite vary according to the mix methods, and the size of the material (from bulk to nanoscale), the structure of magnetite was originally determined in 1995 using X-ray diffraction (**Figure 1**). When magnetite was initially identified in the region, its structure was first discovered using X-ray diffraction. Magnetite is naturally ferromagnetic and has an inverted spinal structure as a result of the alternating lattices of iron (II) and iron (III). The fact that it includes both divalent and trivalent iron sets it apart from other iron oxide [2].

## 2. Classification of iron and iron oxide nanoparticles

$Fe_2O_3$ is primarily used in the fields of sensors, coatings, UV radiation blocking, and coloring. $Fe_2O_3$ nanoparticles are also ideal for exploring nanoparticle polymorphism, magnetic transition, and structural phase transition.

- Alpha $Fe_2O_3$ is an antiferromagnetic material with hexagonal structure corundum, its magnetic momentum is extremely low, and at 1 emu/cc, the transition is seen in alpha $Fe_2O_3$. The alpha structure exhibits mild ferromagnetism in the transition above 260 K, whereas it exhibits anti-ferromagnetism below 960 K. This is a low-temperature transition, and it is dependent on the size of the particles.

- The beta $Fe_2O_3$ is a material that is paramagnetic and has a cubic structure.
- The gamma polymer is a ferromagnetic molecule with a cubic spinal structure. At room temperature, it has a magnetic moment of 430 emu/cc.
- The orthorhombic structure of the epsilon exhibits ferromagnetic molecules (**Figure** 2) [2].

## 2.1 Benefits of green synthesis

The following advantages of green/biological synthesis (**Figure 3**) methods are comparable to those of frequently used physical and chemical synthesis protocols:

a. A method that is safe for the environment and uses no toxic substances.

b. By acting as a capping and reducing agent, the active biological component, which may include an enzyme, lowers the total cost of the manufacturing process.

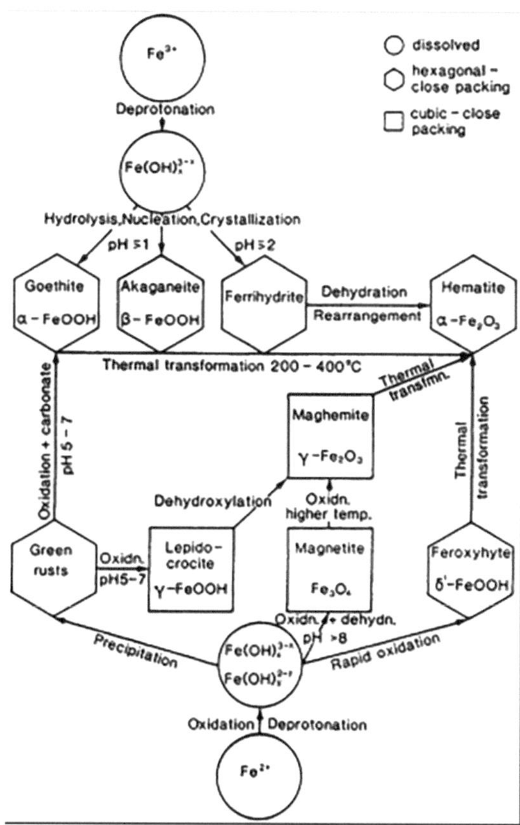


**Figure 2.**
*Formation and transformation of iron and iron oxides nanoparticles [3].*

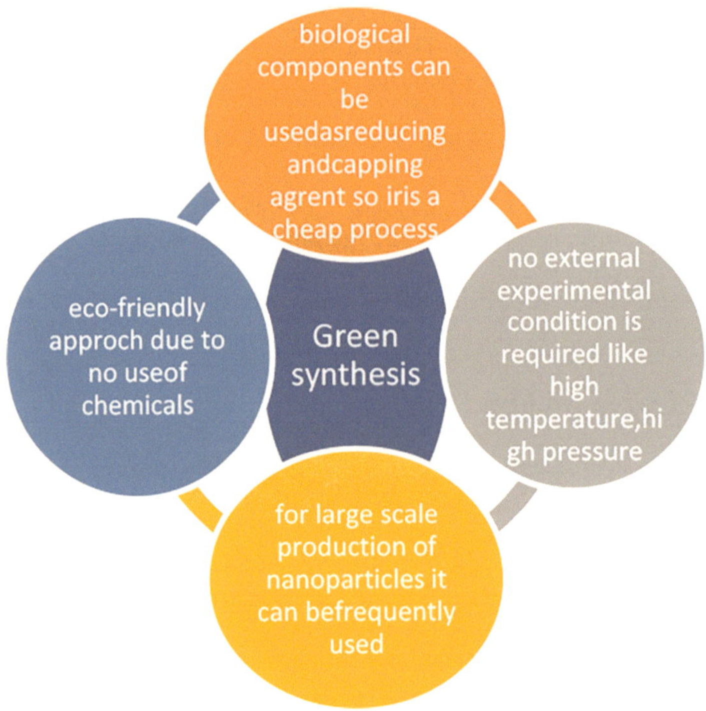


**Figure 3.**
*Diversity of green synthesis.*

c. Even during mass manufacturing and external experimental circumstances like high energy, small nanoparticles can be created [4].

## 3. Synthesis methodology of nanoparticles

Mostly two synthesis methods are used in the procedures, which are following

- Wet chemistry method
- Nonwet chemistry methods.

Precipitation, co-precipitation, hydrothermal, reduction, solvothermal, cryochemical synthesis, and spray/laser paralysis are an example of wet chemistry methods. The last three processes exhibit variation from the surrounding temperature. Cryochemical synthesis takes place below room temperature, and solvothermal synthesis and spray pyrolysis take place above it. Nonwet chemical synthesis includes inert gas condensation, laser ablation, laser-assisted synthesis, and plasma-assisted synthesis [5].

## 4. Mechanism of nanoparticles synthesis

At present, there is no consensus on the mechanism underlying the synthesis of metallic nanoparticles mediated by plant extracts. However, it is widely recognized that secondary metabolites found in various plant components play a critical role in the process. These secondary metabolites include polyphenols, flavonoids, tannic acids, terpenoids, ascorbic acids, carboxylic acids, aldehydes, and amides. Many investigations have employed the IR spectroscopic technique to confirm the existence of numerous reducing sugars, which are often present in plants. Plant extracts containing phytochemicals with appropriate redox characteristics can effectively reduce metal precursors and convert them into matching metallic nanoparticles. According to one hypothesis, radical tannins "R" reduce metal under the influence of pH.

### 4.1 Method

The following methods can be used to initiate the bio-reduction process:

$$FeCl_3 + H_2O \longrightarrow \left[Fe(H_2O)_n\right]^{+3} + H_2O \tag{1}$$

$$R + \left[Fe(H_2O)_n\right]^{+3} + H_2O \longrightarrow \left[Fe(H_2O)_n\right]^{+2} + H^+ + R - OH \tag{2}$$

$$\left[Fe(H_2O)_n\right]^{+2} 2R^- + H^+ + OH^- \longrightarrow Fe^{+2}O^{-2} + O^{-2} + R - H + R - OH \tag{3}$$

The iron-polyphenol complex nanoparticles (Fe-P NPs) structure was produced by eucalyptus leaves, in a different research. The polyphenols found in eucalyptus extract possess the ability to convert $Fe^{3+}$ into $Fe^{2+}$ and at its reduction potential. However, the extract does not entirely convert the $Fe^{2+}$ to zero-valent iron [6]. Due to the presence of polyphenol ligands, $Fe^{2+}$ is firmly stabilized; yet, $Fe^{3+}$polyphenol complexes are quickly formed when $Fe^{2+}$oxidizes in the presence of oxygen, which is known as autoxidation. As a result, a black nano-iron colloid is produced when an iron metal solution reacts with a plant extract. An experiment using the X-ray absorption (XAS) spectroscopy technique indicated that plant polyphenols were located in a globular position and chelated with ferric ion ($Fe^{3+}$). A similar response mechanism was hypothesized for Sage (Salvia). Plant polyphenols can be cross-linked via polyphenol condensation following a reaction with iron chloride ($FeCl_3$) (Eq. (1)) [4, 5, 7].

## 5. Green synthesis of nanoparticles

Compared to microorganisms, plants generate more stable metal nanoparticles and are considered the best candidates for large-scale and quick synthesis. Plants naturally contain a wide range of organic reducing chemicals that are well-suited to the synthesis of nanoparticles. Higher levels of antioxidants found in seeds, fruits, leaves, and stems are masked by the diversity of herbs and plant sources. Therefore, the use of plant-based phytochemicals in the entire synthesis and construction of nanoparticles creates an essential synergy between natural/plant sciences and nanotechnology. This affiliation offers nanotechnology a distinctly "green" perspective, known as green nanotechnology, which can be used without causing significant environmental contamination, setting new benchmarks for highly viable clean and green technology [4, 7]. In contrast to microbe-based synthesis, the creation of nanomaterials from plants is more uniform and faster. Various plant components, including the leaf, stem, seed, and root, are commonly used in the easiest, most affordable, and reproducible method of creating metallic nanoparticles (**Table 1** and **Figure 4**) [27].

| Plant | Material | Parameters Temp (°C) / stirring (rpm) /time ($min^{-1}$) / pH | Size of iron and iron oxide NPs (nm) |
|---|---|---|---|
| *Camellia sinensis* (Leaf) | Fe $(NO_3)_3.9H_2O$ | 75/5000/04 | nZVI (10–20) [8, 9] |
| *Azadirarachta indica* (Leaf) | $FeCl_3.6H_2O$ triple distilled water | 70/5000/4–5 | 9–12 [10] |
| *Punica granatum* (Leaf) | $NH_4Fe(SO_4)_2.6H_2O$<br>$NH_4Fe(SO_4)_2.12H_2O$<br>NaOH | 30/6000/15 | 46.1 [11] |
| *Punica granatum* (Seed) | $FeCl_3.6H_2O$ (99.9%) Distilled water | 70/15000 /10 | 40–50 [11] |
| *Platanus orientalis* (Leaf) | $Fe(NO_3)_3.9H_2O$ (98%)<br>Dabouraud dextrose agar and potatoes agar<br>Hexahit<br>Double distilled $H_2O$ | 25/ 6000/20 | 38 [12, 13] |
| *Daphne mezereum* (Leaf) | $FeCl_3.6H_2O$Conc.HCl<br>Deionized water(DI) | 50/6000/25 | 9–10 [14] |
| *Lagenaria Siceraria* (Leaf) | $FeCl_3.6H_2O$ (98%) Distilled water | 40/4000/60 | 30–100 [15] |
| *Spinancia oleracea* (Leaf) | $FeCl_3.6H_2O$ Deionized water (DI) | 60/5000/30 | 10–100 [6, 16, 17] |
| *Avicennia marina* (Flower) | Ferric chloride ($FeCl_3$)<br>Triple distilled de-ionized $H_2O$ | 125/10,000/30 | 30–100 [18] |
| *Cupressus sempervirens* (Leaf) | $FeCl_3.6H_2O$<br>The HCl solution, Ultrapure water, Al-foil | 50/5000/10 | 9–31 [19] |
| *Vaccinium sect. Syanococcus* (Leaf) | $FeCl_3.6H_2O$<br>Distilled water<br>$CH_3COCH_3$ | −18/9000 / 30 | 52.4 [20] |
| *Glycosmis mauritiana* (Leaf) | $FeCl_2.4H_2O$ $FeCl_3.6H_2O$<br>DI water NaOH<br>Whatman's filter paper | 80/2000/120 | 100 [21] |
| *Syzygium aromaticum* (Bud) | Distilled water, Whatman's filter paper NaOH<br>$Fe(NO_3)_3$ | 25/13,000/20 /pH 6 | 216–268 [22] |
| *Camellia sinensis assamica* (Leaf) | $FeCl_3.6H_2O$<br>Distilled water<br>Filter paper<br>Mortar and pestle Ethanol | 65/10000/20 /pH 7 | 40–50 [23] |
| *Rosa indica* (Leaf) | $FeCl_3.6H_2O$<br>Deionized water<br>Filter paper | 50–60 /5000/5 | 51 [24] |
| *Terminalia Chebula* (Fruit) | $FeSO_4.7H_2O$<br>Distilled water<br>Ethanol | — | Less than 80 [25] |
| *Rumex acetosa* (Leaf) | $FeCl_3.6H_2O$ (99.9%) Distilled water $FeSO_4.7H_2O$ NaOH | 500/4000/15 | 10–40 [23] |

| Plant | Material | Parameters Temp (°C) / stirring (rpm) /time ($min^{-1}$) / pH | Size of iron and iron oxide NPs (nm) |
|---|---|---|---|
| *Bellissima* (Fruit) | Ethanol<br>Distilled water, $FeCl_3$ | — | 22.3 [26] |

**Table 1.**
*Different plant extracts for the green synthesis of iron and iron oxides.*

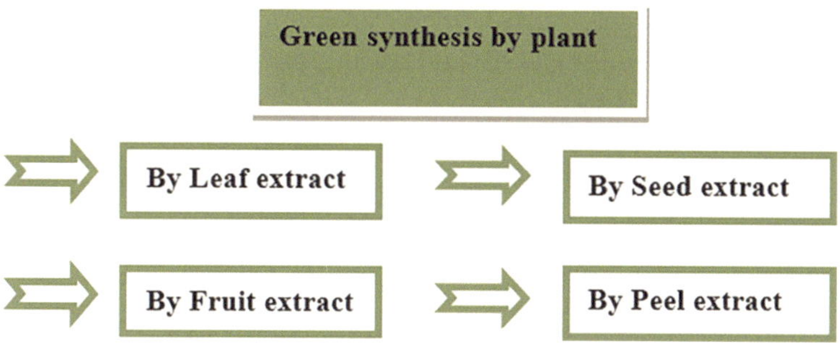


**Figure 4.**
*Flow sheet of green synthesis.*

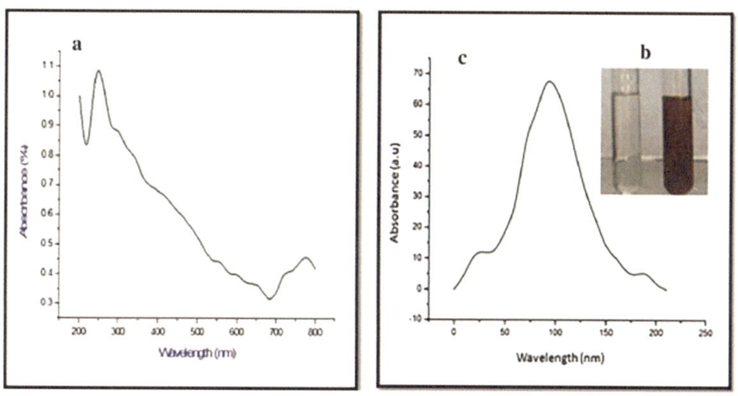


**Figure 5.**
*The UV–visible spectrum of FeNPs [28].*

## 6. Characterization of iron and iron oxides nanoparticles

### 6.1 UV: Visible spectroscopy

The UV-Visible spectroscopy of iron nano.
Particles is in the absorption peak of 216–265 nm (**Figure 5**).

### 6.2 Fourier transform infrared (FTIR) spectroscopy

Peaks at 991,1070, 1631, 2800, and 3467 $cm^{-1}$ in the Fourier transform infrared (FTIR) spectrum of green-synthesized iron nanoparticles (INPs) were caused by the bonds C—O, C=O, C—H, and OH, respectively (**Figure 6**).

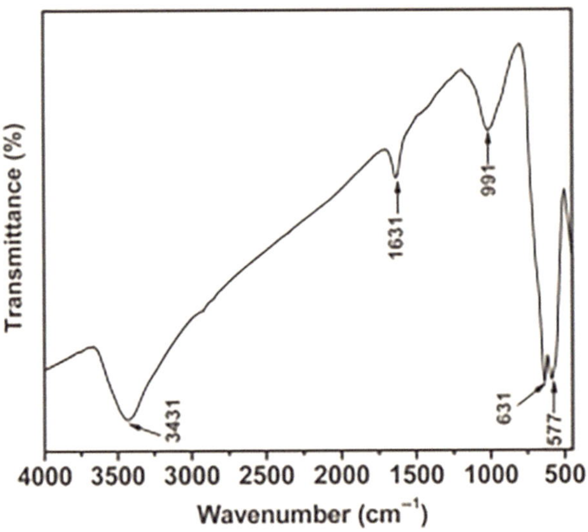


**Figure 6.**
*FTIR spectrum of FeNPs [29].*

## 6.3 X-ray diffraction spectroscopy

The characteristics peak of iron nanoparticles were observed in the XRD pattern at 2θ-values 18.97, 29.81, 35.24, 39.53, and 48.30°. The iron nanoparticles were crystalline in nature. The reduction of metal ions by the extract is clearly showing the iron nanoparticles by X-ray diffraction. By using the Scherer formula we can find out the average particle size of iron oxide nanoparticles [30]. The average size of iron oxide nanoparticles is 45.09 nm (**Figure 7**).

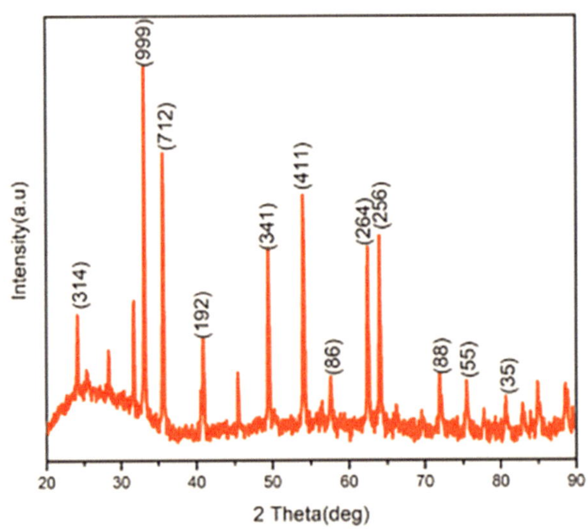


**Figure 7.**
*XRD analysis of FeNPs [31].*

**Figure 8.**
*SEM spectrum of FeNPs [32].*

### 6.4 Scanning electron microscopes

This scanning electron microscope (SEM) is mostly used to identify the structure of the nanomaterial; it can tell us the morphology of the nanoparticles, and the image of iron nanoparticles obtained by the scanning electron microscope looks like a honeycomb. The image of iron nanoparticles in a scanning electron microscope is seen as nonuniform, and its size ranges from 10 to 40 nm with a smooth surface, but they are not uniform (**Figure 8**).

### 6.5 Field emission-scanning electron microscopy and zeta potential

Although there is much disagreement on the size threshold to separate NPs from bulk materials, size is a crucial factor in defining NPs [33–35]. The particles were suspended in water that had a viscosity of 0.887, a refractive index of 1.33, and a dielectric constant of 78.5. The zeta potential reflects the potential difference between the electrophoretically mobile particles EDL (electric double layer) and the layer of dispersant surrounding them at the sliding plane. The potential at the slipping/shear plane of a colloid particle moving through an electric field is also known as electrokinetic potential. The particle size distribution is a result (**Tables 2–4** and **Figure 9**).

| Plants | FTIR ($cm^{-1}$) |
|---|---|
| *Azadirachta indica* (leaf extract) | O—H at 3268.9 (stretching) and COOH at 1636.3 (stretching) [10] |
| *P.granatum* (seed extract) | C=C at 1500 C=N at 1640 C=O at 1720 N—H(amine salt) at 3000 N—H at 3400 [11] |
| *P.orientalis* | C—H at 3196(stretching) H—C—H at 1315(bending) C—O at 1410 C—C at 1000–1450 Fe—O—Fe at 663,462,426 [12, 13] |
| *D.mezereum* | Fe—O at 516<br>O—H of carboxylic acid at 3448<br>C=C at 1635 (ring stretching) [14] |

| Plants | FTIR ($cm^{-1}$) |
|---|---|
| *L.siceraria* | O—H bond at 3354<br>C=O at 1701<br>Inorganic compound at 624 [15] |
| *Spinancia oleracea* (leaf extract) | OH at 3500–3200<br>C—H at 2800<br>C=O at 1658<br>C=C at 1540,1402<br>Fe NPs at 670, 612.63 [6, 16, 17] |
| *A. marina* | M—O bond at 400–450<br>Fe—O at 618 and 467<br>Aromatic comp. at 2923,2853<br>C—H bonds at 3422<br>O—H group at 1630 [18] |
| *G. mauritiana* | OH at 3334.7 (stretching) C≡N at 2115.9 (stretching)<br>(—C=O) at 1632 [21] |
| *Camellia sinensis assamica* | C—O—C at 1064<br>C=C at 1616 (stretching)<br>O—H at 3355<br>C=C at 1608 (ring stretching) [23] |
| *Murraya koenigii* | OH at 3454 (stretching)<br>C—H at 2427 (stretching)<br>C=O at 1767 ester group (stretching)<br>The amino acid at 1637<br>Germinal $CH_3$ group at 1384<br>Fe—O at 831, 537, and 459 (stretching) [36, 37] |
| *Urtica dioica* | C—O at 1070<br>C=O at 1636<br>C—H at 2800<br>OH at 3467 [23, 37] |
| *Pinus eldarica* | O—H at 3420.97<br>C=O at 1616.05<br>C—O at 1069.05<br>C-H at 2918<br>Fe—O at 640 and 450 [7, 36] |

**Table 2.**
*FTIR analysis of different plant extract.*

### 6.6 Application of nanoparticles

Magnetic nanoparticles have a variety of uses, including in biomedicine, healthcare, agriculture, food production, environmental management systems, energy, textiles, electronics, building materials, machines, etc. (**Figure 10**).

## 7. Conclusion

This study focuses on the engineering of iron nanoparticles using several environmentally friendly techniques and their potential to remove contaminants from the environment (**Table 5**). To some extent, efforts are made to emphasize the many

| Plants Extracts | SEM (nm) | TEM (nm) |
|---|---|---|
| *Camellia sinensis* | 16 | 14 [8, 9] |
| *Punica granatum* seed | 10–40 | 38 [11] |
| *P.orientalis* | — | 9.2 [12, 13] |
| *D.mezerium* | — | 20–50 [14] |
| *S. oleracea* | 30–100 | Below 100 [6, 16, 17] |
| *A. marina* | Average 80.2 | Average ~ 52.4 [18] |
| *Vaccinium sect. Syanococcus* | 58–79 | 20 [20] |
| *G. mauritiana* | 50–80 | 10–20 [21] |
| *C.sinensisassamica* | 15–45 | Size depends on the Polyphone concentration ranges from 40 to 200 [23] |
| *Rosa indica* | 61 | 51 [24] |
| *Murrayakoenigii* | 500 | 7–15 [37] |
| *M. indica* | 0.075–6.5 μm | 11.5 [15] |
| *Laurus nobilis* | 200 | 60 [8] |

**Table 3.**
*TEM and SEM value of Fe and Fe—O nanoparticles of different plants.*

| Plants | XRD Pattern result |
|---|---|
| *Punica granatum* (Pomegranate seed extract) [8]<br>Kingdom: Plantae<br>Family: Punica | The sharp peaks at 18.38°, 30.18°, 35.55°, 43.25°, 53.60°, 57.11°, 62.76°, 71.30°, and 74.50° 141 corresponded to the crystal planes of 142 (111), (220), (311), (400), (422), (511), (440), (620), and (533). This demonstrated that the majority of the 143 IONPs synthesized were $Fe_3O_4$. |
| *P.orientalis* [JCPDS 00–003-0863]<br>Family: Platanaceae<br>kingdom: Plantae<br>Order: Proteales<br>Rank: Species | The XRD pattern of Fe—O revealed the presence of three peak of $\alpha Fe_2O_3$,$\gamma Fe_2O_3$ whoes hkl value is (012),(410) and (221) respectively. |
| *A. marina* [PDF No.: 46–1312 JCPDS]<br>Kingdom: Plantae<br>Family: Acanthaceae<br>Genus: Avicennia<br>Species: A. marina Order: Lamiales | The XRD pattern of FeO-NPs revealed the presence of FeO at five peaks:<br>36.42°,42.24°,61.29°,73.24°, and 77.07°, which correspond to the (111), (200), (220), (311), and (222) planes of FeO, respectively. |
| *L.siceraria* [JCPDS card No. 39–1346]<br>Kingdom: Plantae<br>Order: Cucurbitales<br>Family: Cucurbitaceae<br>Genus: Lagenaria<br>Species: Lagenaria siceraria | It is discovered that the exits strong diffraction peaks with 2 values of 28.26° and 32.28°, which correlate to the hkl value of (220),(222), denoting the crystalline phase of $Fe_3O_4$-NPs. |
| *A.indica*<br>[JCPDS file No.19–0629]<br>Kingdom: Plantae<br>Order: Sapindales<br>Family: Meliaceae | The $Fe_3O_4$ XRD pattern revealed six typical diffraction peaks at planes $2\theta$ = (220) at 30.3° (311) at 35.6°, (400) at 43.3°, (422) at 53.2°, (511) at 57.1°, and (440) at 62.8°. |

| Plants | XRD Pattern result |
|---|---|
| *G. mauritiana* [JCPDS NO. 89–8104]<br>Kingdom: Plantae<br>Class: Rosidaceae<br>Order: Sapindales<br>Family: Rutaceae<br>Genus: Glycosmis<br>Species: mauritiana | The significant strong characteristic peaks of iron oxide particles are found at $2\theta = 24.14°$, 33.14°, 35.61°, 40.84°, 49.45°, 54.06°, 62.42°, and 64.00°, which correspond to iron oxide amorphous structures (220), (311), (400), (442), (511), and (440). |
| *M.indica* [(α-$Fe_2O_3$) JCPDS 87–1164] [(γ-$Fe_2O_3$) JCPDS 39–1346]<br>Common name: mango<br>Family name: Anacardiaceae | Hematite (-$Fe_2O_3$) JCPDS 87–1164, and maghemite (-$Fe_2O_3$) JCPDS 39–1346 in gamma phase. –$Fe_2O_3$ appears (in black) in diffracted planes (311) and (220), where -$Fe_2O_3$ appears (in blue) at (012), (104), (110), (113), (024), and (116). |
| *S.aromaticum*<br>[JCPDF card #39–1346]<br>Common name: clove<br>Plant part is taken: buds<br>Family name: Myrtaceae | The $Fe_2O_3$ characteristic XRD peaks suggest a hematite structure and are indexed as (012), (110), (222), (421), (422), and $Fe_3O_4$ is characterized as (111), (311), (222), (400), (422) and (440), indicating the co-existence of iron oxide phases. |

**Table 4.**
*XRD pattern result of the iron and iron oxides from different plants.*

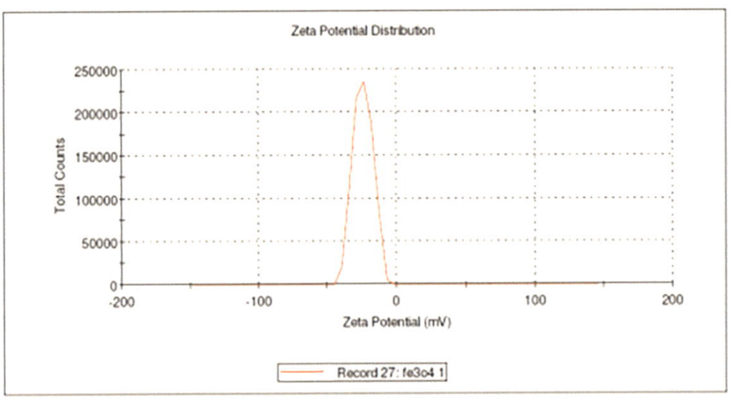


**Figure 9.**
*Zeta potential spectrum of FeNPs [38].*

environmentally friendly agents for the production of iron nanoparticles, including polymers, amino acids, bacteria, fungi, plant extracts, etc. This review also covers the relationship between particle size, morphology, and other aspects and the characteristics of materials, processes, and protocols. According to the literature, several plants and components linked to plants have been used to easily synthesize iron nanoparticles, which have proven to be effective catalysts for a variety of environmental applications. As a result, plant materials appear increasingly viable as producers of iron nanoparticles because of their environmental benefits and high economic worth. To better understand the phytochemistry involved in the creation of iron nanoparticles, which has not yet been thoroughly defined. Additional study is required to examine more nearby and easily accessible resources for the production of iron nanoparticles in order to accomplish the sustainability of nanomaterial synthesis. Any novel methodology must first comprehend the biochemical processes involved in

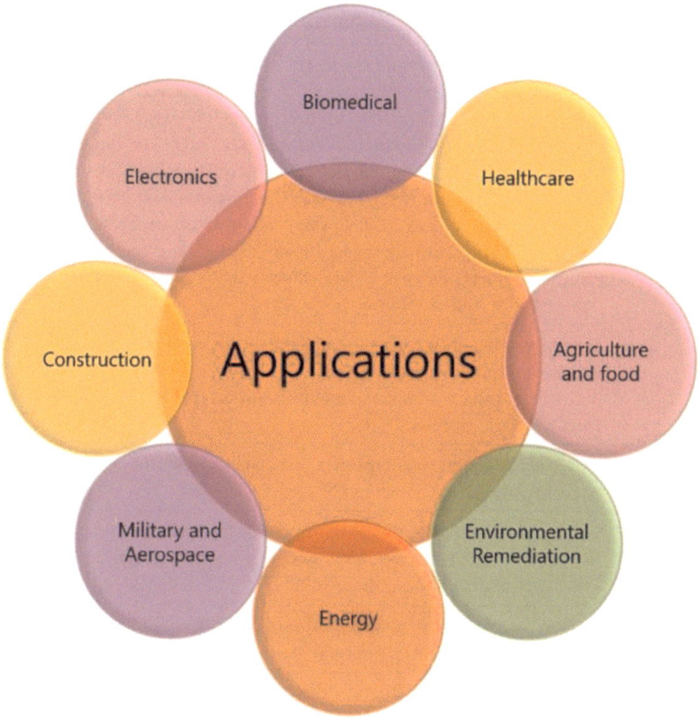


**Figure 10.**
*Application of iron and iron oxides nanoparticles.*

| Field | Application of iron Nanoparticles |
|---|---|
| Biomedical | Cellular therapies (cell labeling/tissue repair/cell separation and handling/purifying cell populations/magneto-reception, diseases of the musculoskeletal system/severe inflammation, disability, and pain are all possible uses for magnetic nanoparticles (especially those coated with liposomes) for drug delivery [39]. |
| Health care | Nanophotothermolysis with pulsed lasers for the treatment of cancer, hepatitis B virus, respiratory syncytial virus, influenza virus, antiviral agents against HIV-1, Monkeypox virus, herpes simplex virus type 1, and Tacaribe virus; delivering antigens for a vaccine; nanoscale biosensors and imaging; nanocoatings on surfaces; implants; nanocarrier for vaccination; antimicrobial activities; SLN in drug delivery and research; delivering antigens for a particular disease into the bloodstream; preventing aging of the skin [39] |
| Agriculture and food | Crop yields are increased by nano-based products (nano-fertilizers, nano-fungicides, and nano-pesticides), designed NPs, and CNTs; pyrite NPs are employed as a seed treatment for different plants before the seeds are sown, and more leaf quantity, greater leaf size, and higher biomass. Improved starch breakdown after storage. This suggests that iron pyrite nanoparticles might be created as a commercial seed treatment product (pro-fertilizer). As no NPs are introduced into the soil throughout the procedure, the method is secure. Fewer doses are needed than with chemical fertilizers. There are no negative impacts on plant development. Nanoporous membranes, gene transfer (crop enhancement), nano-composites, nanosensors, nanofood, encapsulation, food packaging, nanocoatings, and precision farming [39]. |

| Field | Application of iron Nanoparticles |
|---|---|
| Environmental Remediation | Detection, monitoring, and treatment of pollution. Treatment of wastewater (adsorption, membrane filtration, and permeable reactive barriers). Palladium (Pd), climate change (carbon capture), synthetic leaves for $CO_2$ sequestration, mineral carbonation, biomimetic carbonation, $N_2O$ breakdown, and methane combustion are some examples of catalyst coatings. Enhances manufacturing processes (efficiency, waste reduction), dematerialization (reduction in material quantity), sensing (pollutant sensors, nanoporous membranes, chemical and bionanosensors, nanowire sensor for explosives), and energy (heat distribution, for example, ceramic-like materials that sufficiently provide the structure's dependability and durability) [39]. |
| Energy | The conversion of waste heat from computers, cars, homes, power plants, etc. into usable electrical power using photovoltaic film coatings, more efficient fuel production and consumption, fuel cells, prototype batteries, aerogels, and thermoelectric materials [39]. |
| Military and aerospace | Smart materials, nano-composites, nanocoatings, sensors, electronics, fuel additives, and energy devices [39]. |
| Construction | Smart materials, concrete additives, nanoscale sensors, nano-composites, and nanocoatings. In order to color concrete, brick, tile, and other building materials, iron oxide pigments are utilized [39]. |
| Electronics | Printed electronics, carbon nanotubes, nanowires, NEMS, spintronics, and quantum dots [39] are examples of emerging technologies. |

**Table 5.**
*Application of nanoparticles in different fields.*

nanoparticle creation, and any solution must be cost-effective relative to traditional approaches. Utilizing local resources can help keep costs down in the long run as their development. A clearer explanation of biomolecules and their function in modulating the creation of nanoparticles will be provided in future studies with a more in-depth investigation. The objective is to alter the rate of synthesis and boost the stability of nanoparticles. Additionally, research should be done to boost the reactivity of iron nanoparticles during manufacture in order to accelerate the breakdown of environmental pollutants with the fewest possible eco-toxicological effects. Few investigations have confirmed that biosynthesized nanoparticles are less hazardous than manufactured nanoparticles. Additionally, a thorough risk evaluation of green-fabricated Fe NPs should be carried out, taking into account the nanoparticles kinetics, fate, transport, aggregation, and dissolution during processing. A variety of biochemical or functionalized nanoparticles can be produced using the green nanotechnology processes described in this paper as a solid foundation. These new products could be used in the environmental restoration sectors.

## Conflict of interest

The authors declare no conflict of interest.

## Author details

Shaista Ali*, Aliya Zahid and Syeda Taskeen Shahid
Department of Chemistry, Government College University Lahore, Lahore, Pakistan

*Address all correspondence to: shaista.ali.19@gmail.com; shaista.ali@gcu.edu.pk

IntechOpen

© 2023 The Author(s). Licensee IntechOpen. This chapter is distributed under the terms of the Creative Commons Attribution License (http://creativecommons.org/licenses/by/3.0), which permits unrestricted use, distribution, and reproduction in any medium, provided the original work is properly cited.

## References

[1] Alagarasi A. Chapter 1. In: Introduction to Nanomaterials. 2011. p. 76. Available from: https://www.researchgate.net/publication/259118068

[2] Rukhsar M, Ahmad Z, Rauf A, Zeb H, Ur-Rehman M, Hemeg HA. An overview of iron oxide ($Fe_3O_4$) nanoparticles: From synthetic strategies, characterization to antibacterial and anticancer applications. Crystals. 2022;**12**(12):1809. DOI: 10.3390/cryst12121809

[3] Hock S. Precipitation of Hematite and Recovery of Hydrochloric Acid from Aqueous Iron (II, III) Chloride Solutions by Hydrothermal Processing. McGill University; 2009

[4] Fahmy HM, Mohamed FM, Marzouq MH, Mustafa ABED, Alsoudi AM, Ali OA, et al. Review of green methods of iron nanoparticle synthesis and applications. BioNanoScience. 2018;**8**(2): 491-503. DOI: 10.1016/j.jece.2020.104569

[5] Monalisa P, Nayak PL. Eco-friendly green synthesis of iron nanoparticles from various plants and spices extract. International Journal of Plant, Animal a nd Environmental Sciences. 2013;**3**(1): 68-78

[6] Turakhia B, Turakhia P, Shah S. Green synthesis of zero-valent iron nano-particles from Spinaciaoleracea (spinach) and its application in wastewater treatment. Journal of Advance Research in Applied Sciences. 2018;**5**(1):46-51

[7] Kheshtzar R, Berenjian A, Taghizadeh SM, Ghasemi Y, Asad AG, Ebrahiminezhad A. Optimization of reaction parameters for the green synthesis of zero-valent iron nanoparticles using pine tree needles. Green Processing and Synthesis. 2019;**8**(1):846-855. DOI: 10.1515/gps-2019-0055

[8] Mahmoud R, Kotp AA, El-Ela FIA, Farghali AA, Moaty SA, Zahran HY, et al. Green synthesis of iron nanoparticles of clove and green coffee origin with an in vivo hepatoprotective investigation. Journal of Environmental Chemical Engineering. 2021;**9**(6): 106320. DOI: 10.1016/j.jece.2021.106320

[9] Huang L, Weng X, Chen Z, Megharaj M, Naidu R. Green synthesis of iron nanoparticles by various tea extracts: A comparative study of the reactivity. Spectrochimica Acta Part A: Molecular and Biomolecular Spectroscopy. 2014; **130**:295-301. DOI: 10.1016/j.saa.2014.04.037

[10] Zambri NDS, Taib NI, Abdul Latif F, Mohamed Z. Utilization of neem leaf extract on the biosynthesis of iron oxide nanoparticles. Molecules. 2019;**24**(20): 3803. DOI: 10.3390/molecules24203803

[11] Bibi I, Nazar N, Ata S, Sultan M, Ali A, Abbas A, et al. Green synthesis of iron oxide nanoparticles using pomegranate seeds extract and photocatalytic activity evaluation for the degradation of textile dye. Journal of Materials Research and Technology. 2019;**8**(6):6115-6124. DOI: 10.1016/j.jmrt.2019.10.006

[12] García DG, Garzón-Romero C, Salazar MA, Lagos KJ, Campaña KO, Debut A, et al. Bioinspired synthesis of magnetic nanoparticles based on iron oxides using Orange waste and their application as photo-activated antibacterial agents. International Journal of Molecular Sciences. 2023;**24**: 4770. DOI: 10.3390/ijms24054770

[13] Devi HS, Boda MA, Shah MA, Parveen S, Wani AH. Green synthesis of

iron oxide nanoparticles using Platanusorientalis leaf extract for antifungal activity. Green Processing and Synthesis. 2019;**8**(1):38-45. DOI: 10.1515/gps-2017-0145

[14] Beheshtkhoo N, Kouhbanani MAJ, Savardashtaki A, Amani AM, Taghizadeh S. Green synthesis of iron oxide nanoparticles by aqueous leaf extract of Daphne mezereum as a novel dye removing material. Applied Physics A. 2018;**124**:1-7. DOI: 10.1007/s00339-018-1782-3

[15] Kanagasubbulakshmi S, Kadirvelu K. Green synthesis of iron oxide nanoparticles using Lagenariasiceraria and evaluation of its antimicrobial activity. Defence Life Science Journal. 2017;**2**(4):422-427. DOI: 10.14429/dlsj.2.12277

[16] Bunea A, Andjelkovic M, Socaciu C, Bobis O, Neacsu M, Verhé R, et al. Total and individual carotenoids and phenolic acids content in fresh, refrigerated and processed spinach (Spinaciaoleracea L.). Food Chemistry. 2008;**108**(2):649-656. DOI: 10.1016/j.foodchem.2007.11.056

[17] Fayyaz B, Zahra MB, Haider MS. Screening of phenolic compounds from spinach (Spinaciaoleracea), green synthesis of iron-nanoparticles and determination of its anti-microbial effect on Escherichia coli. Research Square. 2022. DOI: 10.1016/j.foodchem.2007.11.056

[18] Karpagavinayagam P, Vedhi C. Green synthesis of iron oxide nanoparticles using Avicennia marina flower extract. Vacuum. 2019;**160**:286-292. DOI: 10.1016/j.vacuum.2018.11.043

[19] Ebrahiminezhad A, Taghizadeh S, Ghasemi Y, Berenjian A. Green synthesized nanoclusters of ultra-small zero valent iron nano-particles as a novel dye removing material. Science of the Total Environment. 2018;**621**:1527-1532. DOI: 10.1016/j.scitotenv.2017.10.076

[20] Manquián-Cerda K, Cruces E, Rubio MA, Reyes C, Arancibia-Miranda N. Preparation of nanoscale iron (oxide, oxyhydroxides and zero-valent) particles derived from blueberries: Reactivity, characterization and removal mechanism of arsenate. Ecotoxicology and Environmental Safety. 2017;**145**:69-77. DOI: 10.1016/j.ecoenv.2017.07.004

[21] Amutha S, Sridhar S. Green synthesis of magnetic iron oxide nanoparticle using leaves of Glycosmismauritiana and their antibacterial activity against human pathogens. Journal of Innovations in Pharmaceutical and Biological Sciences. 2018;**5**(2):22-26. DOI: 10.1155/2021/8822645

[22] Ben-Arfa BA, Salvado IMM, Ferreira JM, Pullar RC. Clove and cinnamon: Novel anti–oxidant fuels for preparing magnetic iron oxide particles by the sol–gel auto–ignition method. Journal of Alloys and Compounds. 2019;**786**:71-76. DOI: 10.1016/j.jallcom.2019.01.306

[23] Saif S, Tahir A, Chen Y. Green synthesis of iron nanoparticles and their environmental applications and implications. Nanomaterials. 2016;**6**(11):209. DOI: 10.3390/nano6110209

[24] Afsheen S, Tahir MB, Iqbal T, Liaqat A, Abrar M. Green synthesis and characterization of novel iron particles by using different extracts. Journal of Alloys and Compounds. 2018;**732**:935-944. DOI: 10.1016/j.jallcom.2017.10.137

[25] Kumar KM, Mandal BK, Kumar KS, Reddy PS, Sreedhar B. Biobased green method to synthesis palladium and iron nanoparticles using Terminaliachebula

aqueous extract. Spectrochimica Acta Part A: Molecular and Biomolecular Spectroscopy. 2013;**102**:128-133. DOI: 10.1016/j.saa.2012.10.015

[26] Carrapiço A, Martins MR, Caldeira AT, Mirão J, Dias L. Biosynthesis of metal and metal oxide nanoparticles using microbial cultures: Mechanisms, antimicrobial activity and applications to cultural heritage. Microorganisms. 2023; **11**(2):378. DOI: 10.3390/ microorganisms11020378

[27] Herlekar M, Barve S, Kumar R. Plant-mediated green synthesis of iron nanoparticles. Journal of Nanoparticles. 2014:9. Article ID 140614. DOI: 10.1155/ 2014/140614

[28] Murugan K, Dinesh D, Nataraj D, Subramaniam J, Amuthavalli P, Madhavan J, et al. Iron and iron oxide nanoparticles are highly toxic to Culexquinquefasciatus with little non-target effects on larvivorous fishes. Environmental Science and Pollution Research. 2018;**25**:10504-10514. DOI: 10.1007/s11356-017-0313-7

[29] Hwang SW, Umar A, Dar GN, Kim SH, Badran RI. Synthesis and characterization of iron oxide nanoparticles for phenyl hydrazine sensor applications. Sensor Letters. 2014; **12**(1):97-101. DOI: 10.1166/sl.2014.3224

[30] Kamath V, Chandra P, Jeppu GP. Comparative study of using five different leaf extracts in the green synthesis of iron oxide nanoparticles for removal of arsenic from water. International Journal of Phytoremediation. 2020;**22**(12):1278-1294. DOI: 10.1080/15226514.2020.1765139

[31] Balu P, Asharani IV, Thirumalai D. Catalytic degradation of hazardous textile dyes by iron oxide nanoparticles prepared from Raphanussativus leaves' extract: A greener approach. Journal of Materials Science: Materials in Electronics. 2020;**31**:10669-10676. DOI: 10.1007/s10854-020-03616-z

[32] Rajendran S, Abuthahir S, Syedzahirullah S, Vignesh C, Vikrant K, Vignesh R. International journal of chemical concepts green synthesis of Nano iron oxide particles from mild steel. International Journal of Chemical Concepts. 2017;**3**(2):196-200

[33] Desalegn B, Megharaj M, Chen Z, Naidu R. Green synthesis of zero valent iron nanoparticle using mango peel extract and surface characterization using XPS and GC-MS. Heliyon. 2019;**5** (5):e01750. DOI: 10.1016/j.heliyon.2019. e01750

[34] Ruqeishi MS, Mohiuddin T, Al-Saadi LK. Green synthesis of iron oxide nanorods from deciduous Omani mango tree leaves for heavy oil viscosity treatment. Arabian Journal of Chemistry. 2019;**12**(8):4084-4090. DOI: 10.1016/j. arabjc.2016.04.003

[35] Jamzad M, Kamari Bidkorpeh M. Green synthesis of iron oxide nanoparticles by the aqueous extract of Laurusnobilis L. leaves and evaluation of the antimicrobial activity. Journal of Nanostructure in Chemistry. 2020;**10**: 193-201. DOI: 10.1007/s40097-020-00341-1

[36] Ebrahiminezhad A, Zare-Hoseinabadi A, Berenjian A, Ghasemi Y. Green synthesis and characterization of zero-valent iron nanoparticles using stinging nettle (Urticadioica) leaf extract. Green Processing and Synthesis. 2017;**6**(5):469-475. DOI: 10.1515/gps-2016-0133

[37] Samarawickrama KGR, Wijiayapala UGS, Fernando CAN. Green synthesis of iron nanoparticles using curry leaves (Murrayakoenigii) extract. Malaya

Journal of Matematik. 2022;**219**:4309-4317

[38] Chandrasekar N, Kumar K, Balasubramnian KS, Karunamurthy K, Varadharajan R. Facile synthesis of iron oxides, iron-cobalt and zero-valent iron nanoparticles and evaluation of their antimicrobial synthesis, free radical scavengining activity and antioxidant assay. Digest Journal of Nanomaterials and Biostructures (DJNB). 2013;**8**(2): 765-775

[39] Khan I, Saeed K, Khan I. Nanoparticles: Properties, applications and toxicities. Arabian Journal of Chemistry. 2019;**12**(7):908-931. DOI: 10.1016/j.arabjc.2017.05.011

Chapter 2

# Applications and Toxicology of Iron Oxide Nanoparticles

*K. Lakshmanan Palanisamy, Kesavan Vignesh and Nanjappan Karthikeyan*

## Abstract

Iron oxide nanoparticles have been intensively studied in the last decade for their unusual physical and chemical properties owing to their extremely small size, large specific surface area and number of promising applications. In Medical applications including magnetic resonance imaging, cell separation and detection, tissue repair, magnetic hyperthermia and drug delivery, iron oxide nanoparticles (IONPs) have been extensively used due to their remarkable properties, such as superparamagnetism, size and possibility of receiving a biocompatible coating. The development of magnetic iron oxide nanoparticles with improved biocompatible surface engineering to achieve minimal toxicity, for various applications in biomedicine is much more inevitable. In this article Iron oxide and its applications were discussed when it is nano dimension with its nanotoxicology.

**Keywords:** metal oxide, iron oxide, nanoparticles, super paramagnetic, nanotoxicology

## 1. Introduction

The most important metal oxides are presently Titanium oxide ($TiO_2$), Zinc oxide (ZnO), Magnesium oxide (MgO), Copper oxide (CuO), Aluminium oxide ($Al_2O_3$), Manganese oxide ($MnO_2$) and Iron oxide ($Fe_3O_4$ and $Fe_2O_3$). Iron oxide is a range of chemical compounds made up of iron and oxygen. These are naturally occurring; some form in the soil or in chemical deposits in rocks or mountains. Not all oxides are functional to humans, but numerous metal oxide play prominent roles in industry, cosmetics, etc. Iron and its compounds are abundantly available in nature and readily synthesised in the laboratory. The two most commonly studied iron oxides have been magnetite ($Fe_3O_4$) and maghemite ($\gamma$-$Fe_2O_3$).

**Magnetite** ($Fe_3O_4$) is a black colour mineral, containing both Fe (II) and Fe (III) and exhibiting ferromagnetic behaviour. It develops in a variety of species and aids with orienting. Alternative names for magnetite include ferrous ferrite, iron (II, III) oxide, magnetic iron ore, black iron oxide, and iron oxide.

Maghemite (-$Fe_2O_3$) is an structural reddish-brown ferromagnetic mineral that is similar to magnetite but has a cation deficient site. It happens in soils as a weathering byproduct of magnetite or as a byproduct of heating other Fe oxides, typically in the presence of organic materials.

 IntechOpen

IONPs are naturally occurring particles in the environment that are produced during volcanic eruptions and air pollution. Particles of either $Fe_3O_4$ (magnetite) or -$Fe_2O_3$ (maghemite), can be formed as emissions from vehicles, manufacturing, and energy plants, but they can also be specially chemically synthesised for a number of uses. Various methods can be employed in their fabrication such as synthesis by water in oil microemulsion system, co-precipitation reactions in constrained environments, polyol method, flow injection synthesis and sonolysis [1–3].

As high magnetization values are needed for applications like as imaging and therapy, magnetic behaviour is a crucial factor in the design and synthesis of superparamagnetic iron oxide NPs (SPIONs). Although this may be achieved by applying the highest magnetic field permitted in clinical settings, it is also possible to manipulate the reaction conditions during the synthesis processes to produce particles with a large surface area and high magnetic susceptibility. Iron oxides are chemical compounds composed of iron and oxygen. Altogether, there are 16 recognised iron oxides and oxyhydroxides as shown in **Table 1**. The uses of these various oxides and hydroxides are tremendously diverse ranging from pigments in ceramic glaze, to use in thermite.

**Oxides:**

- Iron (II) oxide, wustite (FeO)

- Iron (II, III) oxide, magnetite ($Fe_3O_4$)

| Type of iron oxide | Colour | Chemical formula | Mag. behaviour at room temp. |
|---|---|---|---|
| Goethite | Yellow brown | $\alpha$-FeOOH | Antiferromagnetic |
| Lepidocrocite | Orange | $\beta$-FeOOH | Paramagnetic |
| Akaganeite | Brown to bright yellow | $\gamma$-FeOOH | Paramagnetic |
| Feroxyhyte | Red brown | $\delta$-FeOOH | Ferrimagnetic |
| | Red brown | $\delta'$–FeOOH | Superparamagnetic |
| | | High pressure FeOOH | |
| Ferrihydrate | Reddish brown | $Fe_5HO_8{\cdot}4H_2O$ | Superparamagnetic |
| Bernalite | Greenish white | $Fe(OH)_3$ | Weakly ferromagnetic |
| | | $Fe(OH)_2$ | |
| Schwertmannite | Yellow | $Fe_{16}O_{16}(OH)_y(SO_4){\cdot}nH_2O$ | Paramagnetic |
| Haematite | Red | $\alpha$-$Fe_2O_3$ | Weakly ferromagnetic |
| Magnetite | Black | $Fe_3O_4$ | Ferromagnetic |
| Maghemite | Dark brown | $\gamma$-$Fe_2O_3$ | Ferrimagnetic |
| | | $\beta$-$Fe_2O_3$ | |
| | | $\varepsilon$-$Fe_2O_3$ | |
| Wustite | Black | FeO | Paramagnetic |

**Table 1.**
*Recognised iron oxides in 16 forms.*

- Iron (III) oxide ($Fe_2O_3$)
- Alpha phase, haematite ($\alpha$-$Fe_2O_3$)
- Beta phase, ($\beta$-$Fe_2O_3$)
- Gamma phase, maghemite ($\gamma$-$Fe_2O_3$)
- Epsilon phase, ($\varepsilon$-$Fe_2O_3$)

**Hydroxides:**

- Iron (II) hydroxide ($Fe(OH)_2$)
- Iron (III) hydroxide ($Fe(OH)_3$), (bernalite)

**Oxide/hydroxide:**

- Goethite ($\alpha$-FeOOH)
- Aakaganeite ($\beta$-FeOOH)
- Lepidocrocite ($\gamma$-FeOOH)
- Feroxyhyte ($\varepsilon$-FeOOH)
- Ferrihydrite ($Fe_5HO_8 \cdot 4H_2O$ or $5Fe_2O_3 \cdot 9H_2O$)
- High-pressure FeOOH
- Schwertmannite (ideally $Fe_8O_8(OH)_6(SO) \cdot nH_2O$

With the formula $Fe_2O_3$, iron oxide, often known as ferric oxide, is an inorganic substance. It is one of the three primary forms of iron oxide; the other two are the uncommon iron (II) oxide (FeO) and the naturally occurring iron (II, III) oxide ($Fe_3O_4$) found in the mineral magnetite. $Fe_2O_3$ is the primary source of iron for the steel industry as the mineral haematite. Acids easily destroy the reddish-brown, paramagnetic $Fe_2O_3$. As rust and iron (III) oxide have similar compositions and share a number of features, it might be helpful to refer to rust by this name in some situations. Rust, also known as hydrated ferric oxide, is thought of as an ill-defined substance. Magnetite and maghemite are preferred in biomedicine because they are biocompatible and potentially non-toxic to humans. Iron oxide is easily degradable and therefore useful for in-vivo applications.

Generally, magnetite ($Fe_3O_4$) and maghemite ($\gamma$-$Fe_2O_3$) are the two main forms which have attracted widespread interest in biomedical applications due to their super paramagnetic properties. $Fe_3O_4$ exhibits the larger magnetism than $\gamma$-$Fe_2O_3$ because it is in a more stable form. It is also known as black oxide and contains both divalent and trivalent Fe ions. Maghemite (-$Fe_2O_3$) is a reddish-brown weathering by product of magnetite that is found in soil. The oldest iron oxide known is haematite ($Fe_2O_3$), which is also known as ferric oxide. These iron oxides are incredibly stable at

| Property | Magnetite | Maghemite |
|---|---|---|
| Formula | $Fe_3O_4$ | $\gamma$-$Fe_2O_3$ |
| Lattice parameter (nm) | $a$ = 0.8396 | $a$ = 0.8347 |
| Crystallographic system | Cubic | Cubic or tetrahedral |
| Structural type | Inverse spinel | Defect spinel |
| Type of magnetism | Ferromagnetic or superparamagnetic | Ferromagnetic or superparamagnetic |
| Density ($g/cm^3$) | 5.18 | 4.87 |
| Melting point (°C) | 1583–1597 | — |
| Hardness | 5.5 | 5 |
| Type of magnetism | Ferrimagnetic | Ferrimagnetic |
| Curie temperature (K) | 850 | 820–986 |
| $M_s$ at 300 K ($A$-$m^2/kg$) | 92–100 | 60–80 |
| Standard free energy of formation (kJ/mol) | −1012.6 | −711.1 |

**Table 2.**
*Physical and magnetic properties of iron oxide nanoparticles.*

ambient temperatures and frequently appear as the byproducts of the transformation of other iron oxides. Some of the physical and magnetic properties are presented in **Table 2** [4, 5].

## 2. Magnetic properties of IONPs

The physical property of magnetic materials known as magnetism results from the mobility of electron orbitals or intrinsic spin caused by the existence of unpaired electrons (**Figure 1**). Unpaired electrons are required for iron and some materials containing iron to exhibit magnetic behaviour. Magnetic solids are better understood as a collection of magnetic dipole moments because of the abundance of electrons in materials.

If a magnetic material is positioned in a magnetic field $H$, the individual atomic moments in the material contribute to induce the magnetic flux inside the materials can be written as

$$B = \mu_o (H + M) \tag{1}$$

where $\mu_o$ is the vacuum permeability ($12.566 \times 10^{-7}$ $VsA^{-1}$ $m^{-1}$) and the magnetization $M = m/V$ is the magnetic moment per unit volume, where $m$ is the magnetic moment on a volume $V$ of the material. In the regime, where the magnetization scales linearly with $H$, it is useful to define the magnetic susceptibility ($\chi$) as,

$$M = \chi H \tag{2}$$

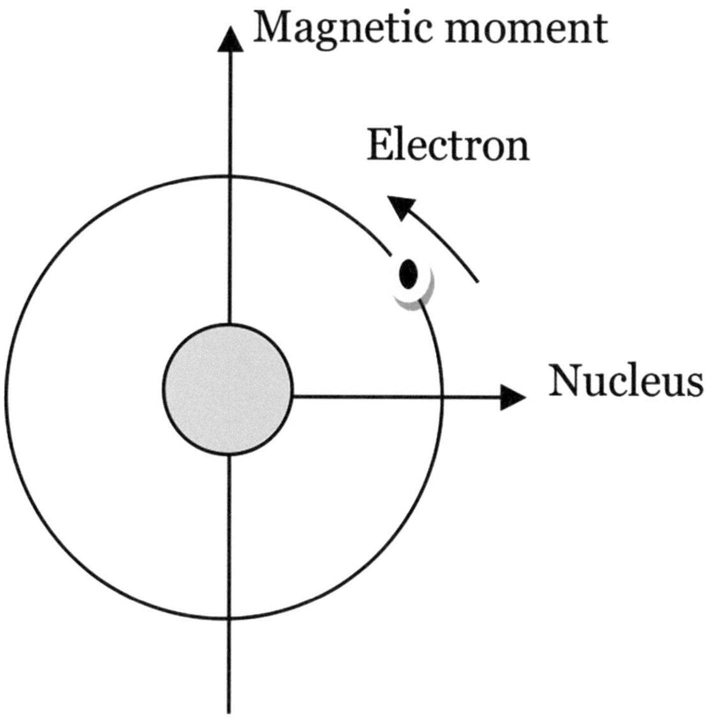


**Figure 1.**
*Magnetic moment of an electron. Source: https://www.embibe.com.*

Fundamentally, there are two types of magnetic measurements for magnetic particles:

1. Magnetization as a function of applied field with temperature (*M-H* loop).

2. Magnetization as a function of temperature at a given magnetic field.

**Figure 2** shows hysteresis loop of magnetic material at constant temperature. Magnetic hysteresis refers to the irreversibility of the magnetization and demagnetization process. The saturated magnetization ($M_s$) is the magnetic moment per unit volume of the material obtained when a sufficiently large magnetic field is applied to remove all domain walls and aligns the magnetization of the sample with the field. Remanent magnetization ($M_r$) is the magnetization that remains after an applied field has been removed. Coercivity ($H_c$) is the applied magnetic field required for reduction of a saturated magnetic material to zero magnetization.

The ZFC-FC curve is obtained by measuring the magnetization, when cooling the sample to the low temperature in the same field as shown in **Figure 3**. In the ZFC and FC measurements, the field must be weak enough in comparison with the anisotropy field to guarantee that the ZFC-FC curve reflects the intrinsic energy barrier distribution.

For the ZFC (Zero Field Cooled) curve, the sample is first cooled in a zero field from a high temperature well above blocking temperature ($T_B$), down to a low temperature much below $T_B$, where nanoparticles are in a ferromagnetic state, where they are in a superparamagnetic state. After that, a magnetic field is provided, and as the

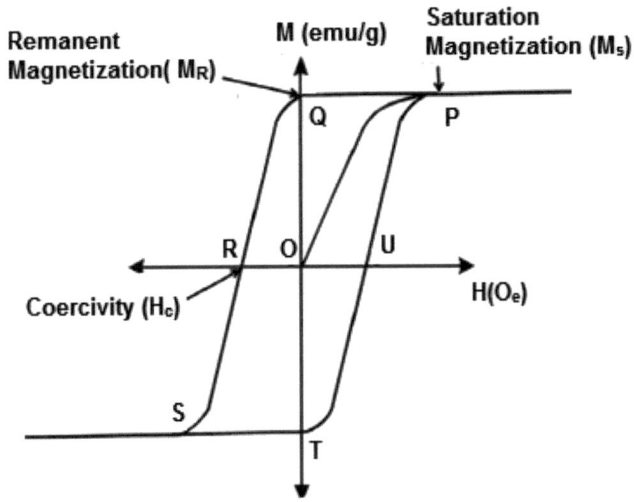


**Figure 2.**
*Hysteresis curve of a ferromagnetic material at constant temperature. Source: http://electrons.wikidot.com.*

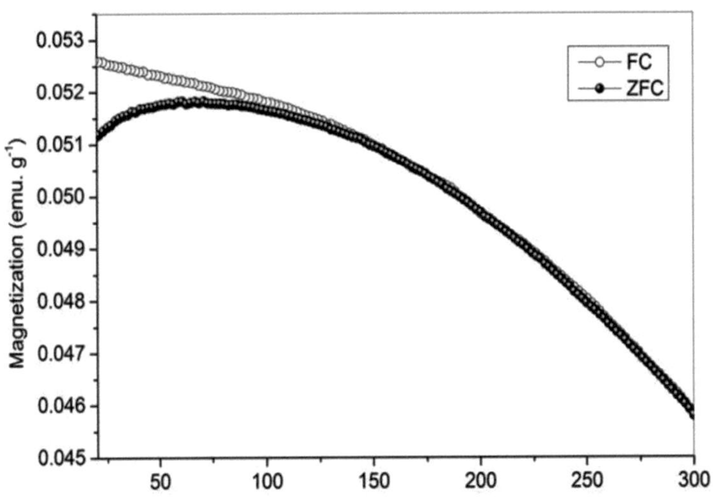


**Figure 3.**
*Typical ZFC-FC magnetization measurement of magnetic material. Source: Materia (Rio J.).*

material is heated to a temperature well over the blocking temperature, the magnetization as a function of temperature is monitored.

## 3. Superparamagnetism

The magnetic behaviour of materials depends on the structure of the material and particularly on its electron configuration. There are several types of magnetism in materials identified as paramagnetism, ferromagnetism, superparamagnetism, antiferromagnetism and ferrimagnetism.

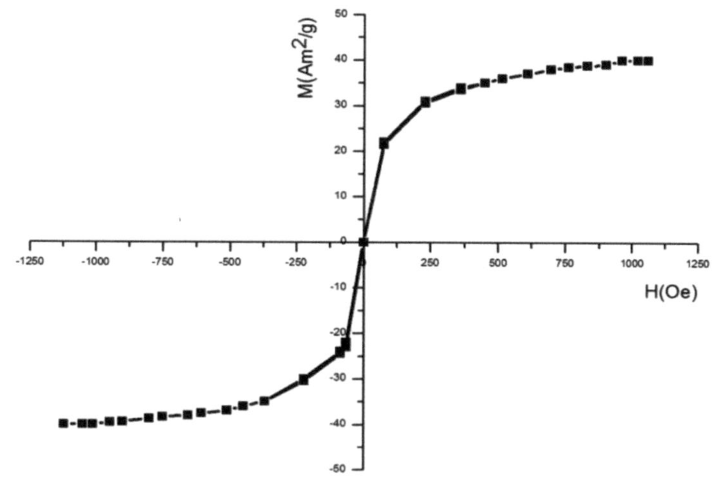


**Figure 4.**
*Hysteresis loop of superparamagnetic material. Source: https://doi.org/10.1063/1.4820992.*

Superparamagnetic materials exhibit non-magnetic moment in the absence of an external magnetic field but can respond to the alignment of the magnetic dipoles along with the external magnetic field. On the other hand, in the presence external magnetic field, they can develop a net magnetic moment. Superparamganetism has a zero intrinsic coercivity, high saturation magnetization, no hysteresis nature like ferromagnetic and no remanence magnetism as presented in **Figure 4**. Generally, superparamagnetism takes place when the material is composed of very small crystallites, i.e. less than 30 nm. Since the particles are very small, no permanent magnetization experiences after the magnetic particles are isolated from the applied magnetic field.

In a superparamagnetic material, Even though magnetic moments are forcing themselves to line up along the field direction, the thermal vibration energy of each particle has a magnitude similar to the magnetic energy, thus the magnetic moments shift their direction arbitrarily as a result. As a result, there is no net magnetic reaction, and the substance acts like a paramagnetic substance. The superparamagnetic materials may still react to an external magnetic field and yet have significant magnetic characteristics with a very high susceptibility. Because of the thinness of the hysteresis loop and lack of remanence or coercively in superparamagnetic materials. The facile resuspension, high surface area, slow sedimentation, stability and dispersion of magnetic force, and lack of magnetic remanence are some benefits of the superparamagnetic particles [6–10]. Subsequently, these advantages make iron oxide crystallites with smaller grain sizes to employ in many applications such as separation, purification, drug delivery device, cancer treatments through hyperthermia and as a contrast agent in Magnetic Resonance Imaging (MRI).

## 4. Applications of IONPs

Iron oxide nanoparticles (IONPs) are fairly a fascinating material have already found in many applications mainly due to large surface to volume ratio and extraordinary magnetic properties. IONPs exhibit sorbent properties, which were successfully tested on removal organic dyes and toxic inorganic metal pollutants from industrial waste water. Moreover, their magnetic properties and ability of the modified surface to selectively bind chemicals show promise of the future industrial interest in magnetically separable sorbents and filtration media.

Maghemite and magnetite NPs have been used in magnetic recording media such as tapes and HDDs. However, current industrial demand for higher recording density have reduced the size of used NPs. High susceptibility and low coercivity of superparamagnetic iron oxide NPs (SPIONs) and their biocompatibility and biodegradability show great potential for applications in biomedicine. In this thesis, some of the applications of IONPs have been discussed briefly which are relevant to our studies.

## 4.1 Biomedical applications of IONPs

Magnetically responsive nano and micro particles have been employed in various areas of biosciences, biotechnology and environmental technology. Biomedicine related applications are mainly based on the utilisation of selected properties, namely magnetic separation, magnetic targeting, hyperthermia, drug delivery, MRI contrast and antibacterial.

Target molecules, cell organelles, and cells can be quickly isolated using magnetic separation from complex biological mixtures and raw samples like blood, bone marrow, tissue homogenates, urine, stools, and other biological materials. Magnetic separation is a very straightforward, quick, efficient, and gentle technique. The chemicals and cells that magnetic separation isolates are often unmodified, pure, and viable. When there is a requirement for large yields of pure and physiologically active chemicals and biological structures on a small scale, gentle test tube magnetic separation is the method of choice.

Hyperthermia treatment is a gifted therapy for cancer where the temperature of the tumour tissue is raised slightly above physiological temperature, i.e. about 42–46°C by artificially. This is based on the fact that when magnetic materials are exposed to an AC magnetic field it generates heat. The heating occurs due to magnetic losses in the form of hysteresis loss. Both bulk magnetic materials and tiny magnetic particles can be employed. Yet due to their advantageous magnetic characteristics in connection to hyperthermia therapy and simplicity of delivery in the target tissues, tiny magnetic particles, especially nanomagnetic materials, have replaced bulk magnetic materials.

Iron oxide particles (IONPs) are routinely used in clinics as a contrast agent for magnetic resonance imaging (MRI). Tracking and monitoring of stem cells in-vivo after transplantation can supply important information for determining the efficacy of stem cell therapy. The most reliable and secure non-invasive method for tracking stem cells in living organisms is thought to be magnetic resonance imaging (MRI) in combination with contrast chemicals. Commercially available superparamagnetic iron oxide nanoparticles (SPIONs) have been used to identify stem cells with the use of transfection agents (TAs).

Magnetic iron oxide nanoparticles attached with drugs has been found that a perspective field of targeted drug delivery where distribution of drug attached with magnetic nanoparticles within the body is manipulated by external magnetic field. A more successful treatment outcome is anticipated if magnetic particles associated with medications can be condensed into the appropriate location by an external magnetic field and then subjected to an AC magnetic field. For the aim of delivering medications, liposomes have been widely used. Liposomes have also been utilised to distribute magnetic particles effectively (known as magnetoliposomes). Targeting of therapeutic agents is necessary for maximum utilisation thereby reducing dose and avoid unwanted adverse effects due to higher doses. Targeting through monoclonal antibody is also used for targeting different drugs/biological molecules to their desired site.

Antibacterial agents are crucial for the food packaging sector, the textile industry, and water purification. The toxicity of organic chemicals used for disinfection is one of its drawbacks. As a result, inorganic disinfectants like metal oxide nanoparticles (NPs) are becoming more popular. The inorganic nanostructured materials with effective surface modification show good antimicrobial inhibition activity. Such improved antibacterial agents locally destroy bacteria, without being toxic to the surrounding tissue. The iron oxide nanoparticles have been widely favoured as antibacterial agents because of low cytotoxicity, biodegradable and reactive surface that can be modified with biocompatible coatings. The mechanism of antimicrobial property of nanoparticle lies with the fact that the large surface area relative to the volume, which effectively covers the microorganisms and reduce oxygen supply for respiration.

### 4.2 Heavy metals and dye removal from water

Water resources become critical important to living things, but there is a major environmental concern due to an increasing pollution from industrial wastewater. Several sectors, including pulp, paper, textile, and plastics, use chemicals and dyes to produce their goods and need a lot of water as well. As a result, harmful substances including heavy metals and chemical compounds infiltrate water. The toxins have negative effects on both the terrestrial and aquatic environments. The use of magnetic IONPs coated with appropriate surfactants to remove pollutants from wastewater shows promise. Magnetic loaded adsorbent materials have gained special attention in water purification due to various advantages such as high separation efficiency, simple manipulation process, fast processing speed, economic and environmental friendly, able to handle concentrated feed, selectivity to particular guest molecules and easy specifically functional modifications.

Municipal wastewater frequently contains copper, which may be harmful to living beings. Techniques including ion extraction, coagulation, or adsorption have been used to remove it, however they have limitations because to their limited sensitivity and cross reactivity. Currently, surface modified IONPs afford an alternative in bioremediation processes for the removal of Cu (II), Cr (VI), Ni etc., from water by adsorption.

The most polluted industry in the world is textiles industry, since it requires a huge amount of two major components, chemicals and water. More than 1000 different chemicals have been used in textiles industry, mostly they are dyes and transfer agents. Natural and synthetic dyes are used in textiles industry for economical and efficiency reasons. It creates problems to the environment, due to the fact that many synthetic dyes are made to be very stable compounds that are resistant to degradation by light, chemicals, and biological processes. Certain synthetic colours used in commerce frequently contain unreportedly huge, complicated structures. Since dye wastewater contains several contaminants, including acid or caustic, dissolved solids, hazardous compounds including heavy metals, and colour, it can be one of industry's biggest problems to discharge. Colour is one of them and is readily identifiable in wastewater since it is very apparent. Moreover, these coloured effluents pose serious environmental risks. Synthetic dyes are present in little amounts, but they are extremely visible and unwanted. The majority of synthetic dyes are also recalcitrant/stable, poisonous, non-degradable, extremely soluble in water, and even carcinogenic.

### 4.3 Corrosion inhibition on mild steel

Corrosion is the destructive attack to metal by chemical or electrochemical reaction problem in mild steel, since which is easily corroded in acidic medium. To guard against corrosive environments, mild steel was coated with a corrosion inhibitor. A corrosion inhibitor is a substance that prevents corrosion by creating a barrier coating of protection, which in turn halts the corrosive reaction. Paint is a substance that stops corrosive media like air and water from coming into direct contact with the metal surface. After curing, it produces a thin, homogenous coating that shields the base metal from corrosion caused by its surroundings. Various metal oxide materials have been used in the formulation of paints. Mixed metal oxides are synthesised from corresponding oxides and are used as pigments in paints which constitute the nanoformulated paints. Nanoformulated paint coatings are being carried out on mild steel plate and the coatings were tested in corrosion atmosphere. The most common corrosion inhibitors are costly, environmentally hazardous organic or inorganic compounds. Mechanistic knowledge of the mechanisms involved in corrosion and inhibition is required for optimal inhibitor selection. As a result, researchers are creating brand-new eco-friendly inhibitors based on natural ingredients and environmentally friendly biopolymers. The organic derivatives found in natural goods with plant origins, such as alkaloids, polysaccharides, pigments, organic acids, and amino acids, can be employed as corrosion inhibitors. Gums and other polysaccharides have been utilised as corrosion inhibitors for metals in salty, acidic, and alkaline conditions.

## 5. Nanotoxicology

Nanotoxicology is the study of the potentially detrimental effects of nanomaterials, in particular nanoparticles, on living creatures. Toxicology is the study of potentially hazardous effects of chemicals on living organisms. Understanding the link between the toxicity of NPs and their physical-chemical characteristics, such as size, shape, reactivity, and material composition, is the main goal of nanotoxicology. There are hundreds of commercially available products using nanotechnology currently on the market including cosmetics, sunscreens, textiles, sport items, veterinary medicines and so on.

The changes in nanoparticle size and structure play an important function on cell toxicity, with rod-shaped or nano-sized IONPs being more toxic than sphere-shaped and microsized particles, respectively. The surface charge of Iron Oxide Nanoparticles could affect cell cytotoxicity and genotoxicity. Absolutely charged IONPs were shown to be more toxic, since they suffer nonspecific interactions and adsorptive endocytosis with the negatively charged cell membrane, thereby increasing their intracellular accumulation and affecting cell membrane integrity.

Nanoparticles are emitted into the environment by primary sources such as natural phenomena, combustion processes or industrial activities during generation and handling of engineered NPs. As NPs are transported through the environment, they can be physically and chemically modified due to interactions with sunlight, water and other environmental substances. The rapid growth of nanotechnology industry and its ever increasing applications will inevitably increase the concentration of nanomaterials in the environment with potential human and environmental exposure as a consequence. The human body is exposed to NPs through four possible routes: inhalation of airborne NPs, ingestion of drinking water or food additives, dermal penetration by skin contact and injection of engineered nanomaterials.

### 5.1 Prevention of toxicity of IONPs

Studies have shown that iron oxide nanoparticles require a biocompatible coating to prevent agglomeration that may lead to toxicity in biological media. The clearance rate, cell toxicity, and cellular response to superparamagnetic iron oxide nanoparticles are significantly influenced by their dimension and surface properties. In general, macrophages are more likely to take up nanoparticles with sizes greater than 200 nm. On the other hand, extravasations and renal clearance both work quickly to eliminate tiny particles with sizes less than 10 nm. This suggests that the best likelihood for a particle to remain in circulation for a longer period of time is between 10 and 200 nm. Amphiphilic coatings are used to increase the circulation duration of superparamagnetic iron oxide nanoparticles from minutes to hours. This increases the targeting potential of a surface-modified contrast agent and prevents a harmful contact between the particles.

Several studies have revealed that the polymer coated nanoparticles barely affect the viability and functionality of cells. A biodegradable polymer coating was used to stabilise the magnetic nanoparticles that Gomez-Lopera et al. created. In their work they were synthesised colloidal nanoparticles that were highly responsive to magnetic field for drug delivery systems. Another study shows that dextran coated superparamagnetic nanoparticles labelled with a cationic peptide had no significant effect on cell viability or the biodistribution of human haematopoietic cells.

In summary, all studies have shown that iron oxide nanoparticles with biological coating very prominent and inevitable in reducing the toxicity to the cells which uptake them. Recently the nanotechnology has been moving towards to develop a new kind of materials without any harmfulness to the society by adopting green synthesis procedures [11–14].

## 6. Conclusion

This article briefly presents the characteristics of nanosized metallic iron oxides, as well as their applications and toxicology. Metal oxides at nanoscale as iron oxides, haematite ($\alpha$-$Fe_2O_3$) and maghemite ($\gamma$-$Fe_2O_3$) have been widely applied in various mankind applications since they are strongly dependent on the size of their particles and surface area. As a result of the reduction of particle size, there is an increase in atoms located on the surface and consequently an increase in the catalytic activity of nanoparticles iron oxidehematite and maghemite.

## Author details

K. Lakshmanan Palanisamy*, Kesavan Vignesh and Nanjappan Karthikeyan
Department of Physics, Sengunthar Engineering College, Tiruchengode, Tamilnadu, India

*Address all correspondence to: klpphysics@gmail.com

IntechOpen

© 2023 The Author(s). Licensee IntechOpen. This chapter is distributed under the terms of the Creative Commons Attribution License (http://creativecommons.org/licenses/by/3.0), which permits unrestricted use, distribution, and reproduction in any medium, provided the original work is properly cited.

## References

[1] Abdel Hameed RS, Al-shafey HI. "Recycling of some plastic waste as green corrosion inhibitors for steel in 1M Phosphoric acid corrosive medium". In: Eurasia 12 Conference, Accepted as Oral Representation. MSAIJ. 2016;**14**(11):417-425

[2] Abboud Y, Abourriche A, Saffaj T, Berrada M, Charrouf M, Bennamara A, et al. The inhibition of mild steel corrosion in acidic medium by 2,20-bis(benzimidazoles). Applied Surface Science. 2006;**252**(23):8178-8184

[3] Abdel-Gaber AM, Abd-El Nabey BA, Khamis E, Abd El-khalek DE. "The novel environmentally friendly plant extract as anti sale and corrosion inhibitor". In: The 24th Annual Conference: Correction Problems in Industry, Egyptian Corrosion Society. 2005. pp. 5-8

[4] Anca C, Ioana M, Vaireanu DI, Iosif L, Carmen L, Simona C. Estimation of inhibition efficiency for carbon steel corrosion in acid media by using natural plant extracts. Revista de Chimie. 2009;**60**(11):1175-1180

[5] Ashassi-Sorkhabi H, Seifzadeh D. The inhibition of steel corrosion in hydrochloric acid solution by juice of Prunus cerasus. International Journal of Electrochemical Science. 2006;**1**:92-98

[6] Kumar A, Vemula PK, Ajayan PM, John AG. Silver-nanoparticle-embedded antimicrobial paints based on vegetable oil. Nature Materials. 2008;**7**:236-241

[7] Abdallah M, Zaafarany IA, Abd El Wanees S, Assi R. Corrosion behavior of nickel electrode in NaOH solution and its inhibition by some natural oils. International Journal of Electrochemical Science. 2014;**9**:1071-1086

[8] Abdallah M, Al Karane SA, Abdel Fattah AA. Inhibition of the corrosion of nickel and its alloys by natural clove oil. Chemical Engineering Communications. 2009;**196**:1406-1416

[9] Poongothai N, Ramachanderan T, Natesan M, Murugavel SC. Corrosion inhibition of mild steel by essential oils in an HCl environment. Materials Performance. 2009;**48**(9):52-56

[10] Rathi VR, Nirmal SD, Kokate SJ. Corrosion study of mild steel, tor steel and CRS steel by weight loss method. Journal of Chemical and Pharmaceutical Research. 2010;**2**(2):97-100

[11] Kumar B. Green synthesis of gold, silver, and iron nanoparticles for the degradation of organic pollutants in wastewater. Journal of Composites Science. 2021;**5**(8):219. DOI: 10.3390/jcs5080219

[12] Gupta A, Sharma V, Sharma K, Kumar V, Choudhary S, Mankotia P, et al. A review of adsorbents for heavy metal decontamination: Growing approach to wastewater treatment. Materials. 2021;**14**(16):4702. DOI: 10.3390/ma14164702

[13] Kumar B, Smita K, Galeas S, Guerrero VH, Debut A, Cumbal L. One-pot biosynthesis of maghemite (γ-Fe2O3) nanoparticles in aqueous extract of *Ficus carica* fruit and their application for antioxidant and 4-Nitrophenol reduction. Waste and Biomass Valorization. 2021;**12**:3575-3587. DOI: 10.1007/s12649-020-01279-9

[14] Kumar B, Smita K, Galeas S, Sharma V, Guerrero VH, Debut A, et al. Characterization and application of biosynthesized iron oxide nanoparticles using Citrus paradisi peel: A sustainable approach. Inorganic Chemistry Communications. 2020;**119**:108116. DOI: 10.1016/j.inoche.2020.108116

Chapter 3

# Biomedical Applications of Superparamagnetic Iron Oxide Nanoparticles (SPIONS) as a Theranostic Agent

*Nancy Jaswal, Purnima Justa, Hemant Kumar, Deepshikha, Krishna, Balaram Pani and Pramod Kumar*

## Abstract

Nowadays, nanoparticles are used in a variety of biological applications where they enhance treatments and laboratory tests. Due to their distinctive properties and minor adverse effects, nanoparticles are being used more specifically for medication delivery, not only in the treatment of cancer but also for other diseases. Magnetic nanoparticles like SPION (superparamagnetic Iron Oxide nanoparticles) are regarded to be the most viable in the midst of these materials. SPION are frequently used in biomedical applications due to their low cost and lack of toxicity. Within the developing field of nanomedicine, superparamagnetic iron oxide nanoparticles (SPION) are basic technological classes that have been widely studied for cancer imaging and treatment. Additionally, SPION employ super paramagnets, which seem to be beneficial for focusing on particular tumor areas within a body. For instance, the superparamagnetic abilities of magnetite ($Fe_3O_4$), which are frequently utilized in delivery of drug, diagnosis and therapy. SPION was envisioned as a tool for the "golden therapeutic era" since it minimized cellular absorption by macrophages, targeted cancer cells preferentially while sparing healthy cells, monitored cancer cells before and after therapy, and controlled drug release. In order to give a concise overview of SPION, there will be focus on their biomedical applications includes hyperthermia (HT), magnetic resonance imaging (MRI), magnetic drug targeting (MDT), gene delivery as well as nanomedicine.

**Keywords:** SPIONs, magnetite, biomedical application, therapeutic, theranostic agent

## 1. Introduction

SPION (Super magnetic iron oxide nanoparticles) are the nanoparticles with a functionalized shell surrounding an iron oxide core. These are iron oxide crystals {magnetite ($Fe_3O_4$) or maghemite ($\gamma$-$Fe_2O_3$)} with a shell that can be altered to

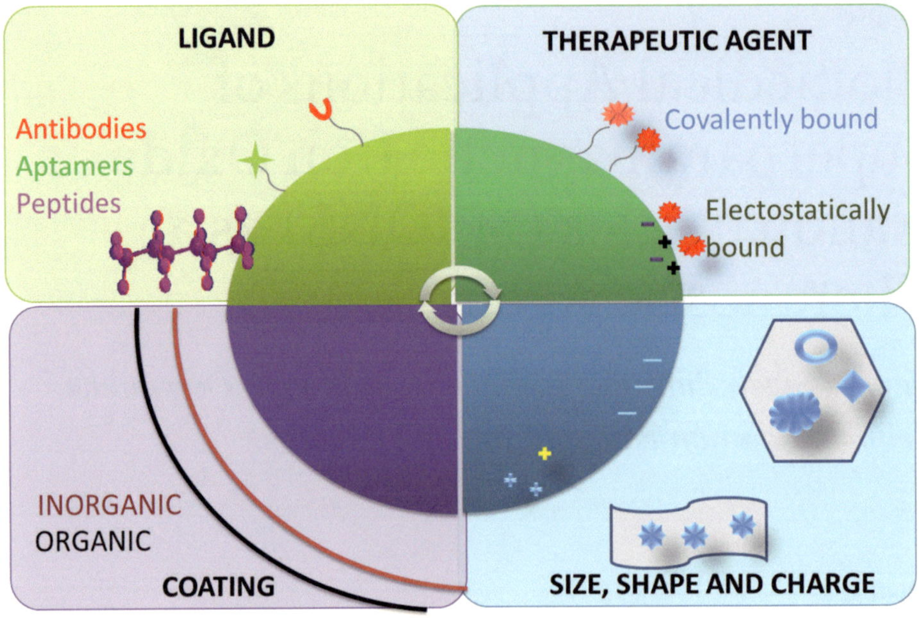


**Figure 1.**
*Showing physiochemical properties of SPIONs [4].*

improve stability in aqueous conditions and to alter the biochemical properties for a wide range of applications in biomedical sectors [1, 2]. The scientific community has recently become very interested in SPION due to their intriguing potential diagnostic and therapeutic applications. They have wide range of biomedical applications such as MRI, dual modality MRI/computed tomography (CT), magnetic fluid hyperthermia (MFH), biosensors, drug delivery as well as bio- separation [3]. SPION are a unique contrast agent for magnetic resonance imaging and a fascinating family of tracers for nuclear medicine due to their modifiable surface and core properties. A contrast agent can accelerate the relaxation rate of water, which is known as relaxation rates, since MRI evaluates the change in magnetic moment of water protons after applying radiofrequency (RF) pulses. In essence, after being excited by RF pulses, protons return to their equilibrium state, which is longitudinal $T_1$ relaxation. Transfer ($T_2$) relaxation is the exchange of the spin angular momentum between the protons (**Figure 1**).

SPION exhibit contrast enhancing behavior, engage with nearby water molecules and speed up the rate at which water protons relax. The $T_1$ relaxation times are sped up by the contrast agents. The regions where SPION are taken up typically have a lower MR a signal intensity, which causes those regions to appear darker in MRI. Transfer ($T_2$) relaxation is the exchange of protons' spin angular momentum. SPION act as a contrast agent, interact with nearby water molecules, and speed up the rate at which water protons relax. The $T_1$ relaxation and/or $T_2$ relaxation times are sped up by the contrast agents. The areas where SPION are taken up typically have a lower MR signal intensity, which causes those areas to appear darker in MRI [5–8]. The current chapter comprehensively reviews the biomedical applications of SPION in theranostic area (**Figure 2**).

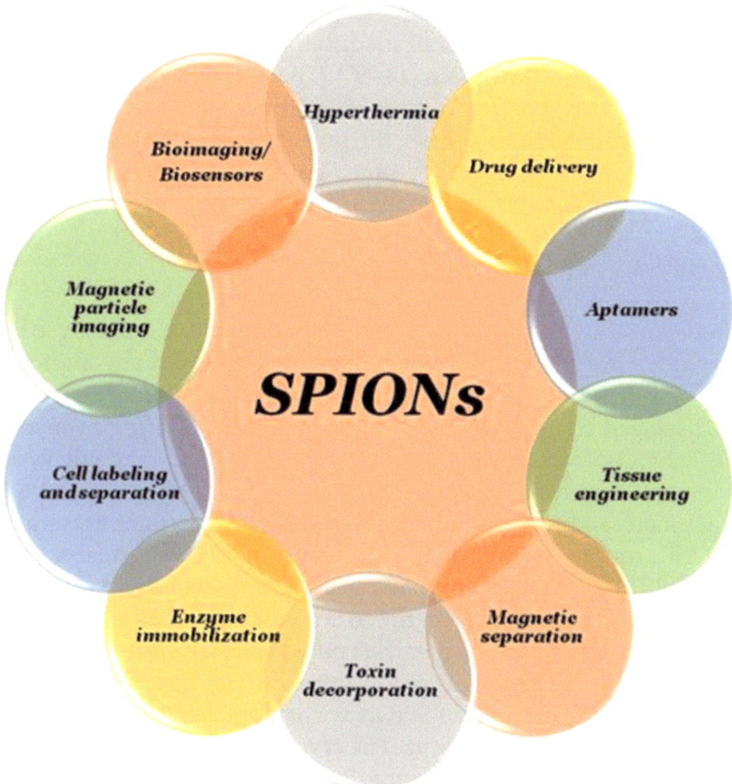


**Figure 2.**
*Showing biomedical applications of SPIONs [9].*

## 2. Synthesis and biomedical applications of SPIONS

### 2.1 Synthesis methods

The morphology as well as composition both affects the magnetic properties of the Iron oxide nanoparticles. In order to ensure control over the particle's size, shape, size distribution as well as crystallinity, the method of synthesis must be carefully chosen. SPION can be created by using as variety of techniques, including chemical, physical and biosynthetic ones [10–13].

#### *2.1.1 Co-precipitation*

The co-precipitation technique, one of the most straightforward and effective synthesis methods, relies on the addition of a weak or strong base to precipitate $Fe^{2+}$ and $Fe^{3+}$ aqueous salt solutions simultaneously. This process is primarily used to synthesize SPION commercially. Several artificial parameters affect the size, shape and the composition of iron oxide nanoparticles, such as $Fe^{2+}/Fe^{3+}$ ratio, temperature, pH, and salt type (chloride, nitrate, sulphate, perchlorate), as well as the kind of base employed (NaOH, $NH_4OH$, $Na_2CO_3$). This technique is one of the most economical ways to make SPION with the right magnetic properties in high degree of polydispersity and little crystallinity. A number of modified versions of this technique have been developed to address these drawbacks [14–16]. These for example, include in vivo co-precipitation in a carboxyl-functionalized polymer matrix, assistance from ultrasound, the use of alkanolamines as base, the preparation of $Fe_3O_4$ nanoparticles under a static magnetic field and at the end co-precipitation of $FeCl_3.6H_2O$, $FeSO_4.7H_2O$, and $Gd(NO_3)_3$ aqueous solutions is done by addition of NaOH [17–20].

### *2.1.2 Microemulsion*

Microemulsion systems are the isotropic dispersion of two immiscible liquids that are thermodynamically stable. Essentially, there are two types of microemulsions: oil-in-water (o/w; normal micelles) and water-in-oil (w/o; reversed micelles). Usually, the dispersed phase acts as a nano/micro-reactor, offering a constrained setting for the initiation and controlled development of nano- and microparticles. Micellar microemulsion systems are created using a variety of amphiphilic surfactants, such as dioctyl sodium dodecyl sulphate (DSS), cetyltrimethylammonium bromide (CTAB), sodium dodecyl sulphate (SDS), and polyethoxylates (such as Tween-20 and -80) [14]. The main benefit of using microemulsion methodology to prepare SPION is the ability to control the nanoparticle size by varying the size of the micelles. Furthermore, a rise in the particle's polydispersity is seen, which is likely caused by the relatively homogenous size of micelles. The typical low and constrained reaction temperature, which leads to SPION with poor crystallinity and low yields, is a drawback of microemulsion synthesis. The crystallinity of the particles produced by this method can be increased by thermal annealing the synthesized iron oxide or maintaining the micelle structure at high reaction temperatures [12, 21, 22].

### *2.1.3 Thermal decomposition*

By thermally decomposing organoiron precursors in high-boiling-point organic solvents with stabilizing surfactants, SPION with excellent size and shape control, narrow size distribution, and good crystallinity can be produced [23, 24]. Oleic acid, oleylamine, fatty acids, and hexadecylamine are examples of amphiphilic surfactants that enable fine-tuning of the nucleation and growth kinetics of the nanoparticles. The presence of the surfactant in the reaction mixture and the high reaction temperatures produce samples with excellent size dispersion and crystallinity. However, because this method involves the synthesis of toxic chemicals like iron pentacarbonyl, chloroform, and hexane, it is not particularly eco-friendly. Furthermore, a further surface modification step is required to produce water-dispersible and biocompatible nanoparticles that are helpful for biomedical applications because the surface of the magnetic nanoparticles has a hydrophobic coating. Control over morphology and nanoparticle size when using the thermal decomposition method to create SPION is highly dependent on reaction time, reaction temperature, and the ratio of precursor to surfactant [25].

### *2.1.4 Sol: gel*

The sol-gel method, which is frequently used for the production of silica-coated SPION, is based on the condensation and hydrolysis of tetraethyl orthosilicate (TEOS) in ethanol and 30% aqueous $H_2O_2$ with $Fe^{3+}$ solutions to create colloidal sols. A 3D iron oxide network is then created by gelling the sol through a chemical

| **Methods of synthesis** | **Reactants used** | **Temperature during reaction (°C)** | **Pressure during reaction(psi)** | **Time taken during synthesis (hrs.)** | **Size distribution (nm)** | **Saturated magnetization at room temp(emu/g)** | **References** |
|---|---|---|---|---|---|---|---|
| Thermal decomposition (liquid phase) | Organometallic compound (Fe(acac)$_3$/ Fe (CO)$_5$) | 100–400 | 14.7 | 20–24 | 4–20 | 20–85 | [26–30] |
| Thermal decomposition (gas phase) | Fe(CO)$_5$, Fe(C2H5)$_2$ (Ferrocene) | 125–1000 | 5.8–14.7 | 0.0041–0.16 | 6–100 | 20–45 | [30–32] |
| Co-precipitation | $FeCl_3$, $FeCl_2$/$FeSO_4$, NaOH/$NH_4OH$ | 25–70 | 14.7 | 0.5–1 | 3–15 | 65–85 | [30, 33] |
| Reduced co-precipitation | $FeCl_3$, $Na_2SO_3$, NaOH/ $NH_4OH$ | 25–70 | 14.7 | 0.5–1 | 3–15 | 65–85 | [30, 34] |
| Hydrothermal | Variable | 80–160 | Up to 2000 | 0.5–48 | 3–15 | 5–75 | [30, 35] |
| Microemulsion | $FeCl_3$, $FeCl_2$/$FeSO_4$, NaOH/$NH_4OH$ | 4–90 | 14.7 | 3–20 | 3–12 | 30–60 | [30, 36] |
| Mechanical | Fe (powder), $FeCl_2$, $FeCL_3$ | Room Temperature | Ambient | 1–48 | 20–100 | 60–146 | [37–39] |
| Sonochemical decomposition | Fe (CO)$_5$, Fe(acac)$_3$, $FeCl_2$, $FeCl_3$ | 0–50 | Ambient | 0.5–8 | 5–30 | 30–80 | [30, 40–42] |

**Table 1.**
*Iron oxide nanoparticle: Synthesis methods and reaction parameters of SPION.*

reaction or solvent removal. After drying and solvent removal, the formed gel needs to be crushed in order to obtain iron oxide nanoparticles. When surfactant is added before gelation, the system's free energy is reduced, which causes the formation of nanoscale iron oxides without the development of a 3D network. A simple method for producing high yields of relatively large and monodisperse nanoparticles in ambient conditions is the sol-gel synthesis technique. But because the sol-gel method is used at room temperature, additional heating is required to produce the desired crystalline structures. The technique also produces contaminated by-products, necessitating post-treatment purification. The parameters that affect the SPION gel's structure and properties are temperature, pH, the solvent being used, and the concentration of salt precursors. TEOS and ammonia concentrations are typically used to adjust the silica shell's thickness (**Table 1**) [26, 27, 43].

## 3. Biomedical applications of SPION

As SPION are mainly used due to their main features such as their size, surface charge, shape, surface stability and biocompatibility. There are many uses of SPION in biomedicine, which are listed below: MRI, drug delivery, hyperthermia, cell labeling and separation, aptamers, biosensors etc. (**Figure 3**) [30].

### 3.1 Drug delivery

The Transmission of a remedial agent to a recommended site in the body is known as drug delivery. Drug delivery to directed disease is a critical step in the treatment of a variety of conditions, including cardiac disease, microbial attack and cancer. For targeted drug delivery of many diseases different kinds of magnetic

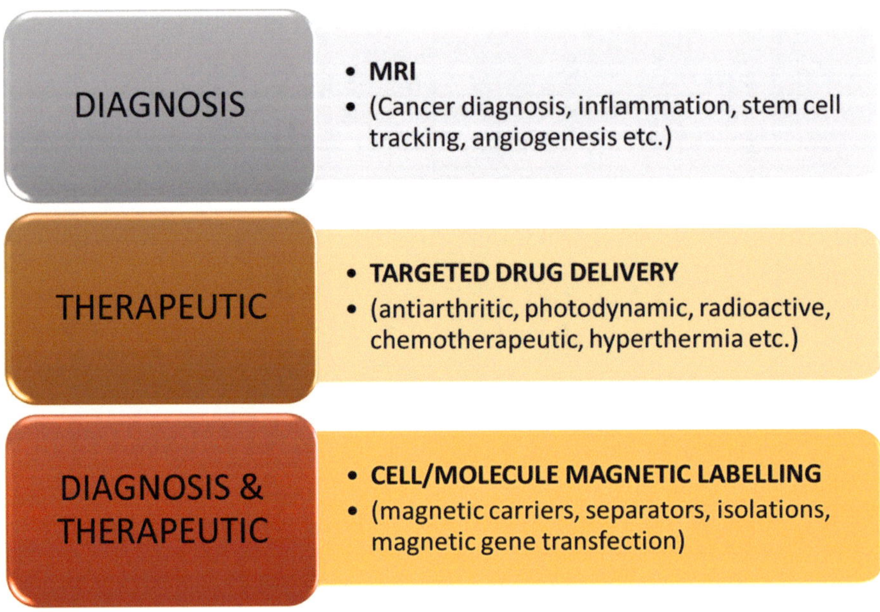


**Figure 3.**
*Showing SPIONs application in biomedicine [30].*

| Functional molecules | Structure of DDS | Application | Results | References |
|---|---|---|---|---|
| Methotrexate | via a self-assembled monolayer PEG it immobilized on the surface of the iron oxide nanoparticles | Through MRI follow-up drug administration to brain tumors | High absorption of targeted nanoparticles, non-toxic methotrexate in vitro, significantly improved contrast | [46] |
| Methotrexate targeting folate receptor | Iron oxide NPs converted with APTES and linked by covalence with Methotrexate | For therapeutic treatment and imaging of cervical tumors and breast | In cells there is greater absorption of nanoparticles targeted with the folate receptor | [47] |
| Folate | Nuclei of SPIONs covered with a mixture of the triblock copolymer methoxy PEG-b-poly (methacrylic acid -co-n-butyl methacrylate)-b-poly (glycerol monomethacrylate) and the folate-conjugated block copolymer folate-PEG-b-poly (glycerol monomethacrylate) loaded with doxorubicin | For the treatment of cancer of cervix | The Targeting strategy improved absorption of nanoparticles and cytotoxicity | [48] |
| Folate | Incorporated inside Pluronic F127 micelles is folate iron oxide | Magnetic resonance imaging and medication administration | Absorption in the KB unit | [49] |
| Murine melanoma antigens, hgp10025–33 | Murine melanoma antigens is brought by SPIONS,hgp10025–33 | Administration of a murine melanoma antigens | Effective adoption of nanoparticles by engineers | [50] |
| Anti-Prostate-specific membrane antigen | A nanocomposite with quantum dots conjugated to the surface and an internally embedded PS matrix that is spherical, high fraction of SPIONs+PLGA+paclitaxel load | Prostate cancer imaging and targeting, medicine storage | Considerable targeting | [51] |
| LHRH (Luteinizing hormone releasing hormone) | PEG and LHRG-coated SPIONs poly(propyleneimine) generation 5 dendrimer siRNA complex | Drug delivery | Improve the internalization of cancer cells and increase the effectiveness of in vitro genetic suppression | [52] |

| Functional molecules | Structure of DDS | Application | Results | References |
|---|---|---|---|---|
| LHRH | Conjugated SPIONs with LHRH | Treatment for human breast cancer tumors and metastases | The targeted nanoparticles had an accumulation a 12-multiplied by the accumulation of breast and lung metastases in vivo | [53] |

**Table 2.**
*Showing use of SPION in drug carrier and preclinical studies.*

nanoparticles are used. For example, in detection lesions in the liver tumor at a 23 mm level. SPION have been used by clinical imaging. For the treatment of MCF-5 breast cancer, drug administration and MRI monitoring have also been studied [39, 44, 45]. In the whole body, chemotherapy causes serious damage, so it is important to focus selectively on cancer cells to create targeted therapies. For higher doses of drugs, the targeting strategy should also increase the effectiveness of drugs without being limited by the harm to healthy tissue. For the SPION-based drug delivery system to be effective, many important criteria must be considered.in order to link targeting units, the carrier should provide a functional group and also provide the delivery system with suitable hydrophily, so that it can easily disperse into aqueous environment (**Table 2**) [8].

### 3.2 Bioimaging

In bioimaging and clinical purpose magnetic nanoparticles show great scientific interest because of their unique properties. At molecular and cellular level magnetic resonance imaging (MRI) is one of the strongest techniques used in biomedical imaging. In magnetic resonance imaging (MRI), many magnetic nanoparticles are used as a contrast medium, such as SPION, the core, magnetic gel nanoparticles and iron oxide. In these SPION demonstrate an important role in differentiating pathogenic and healthy cells. Because of the high resolution and capability of 3D imaging MRI provides information about soft tissue as well as for the detection of intra-tumor cancerous tissue. SPION and magnetic nanoparticles of iron oxide have aroused major interest in the administration of cancer drugs over the last decades [39].

### 3.3 Cell labeling and separation

There is numerous SPION cell labelling techniques that can be applied, each with a different goal in mind. There are three main categories that they fall under:

i. *In-vitro* methods (such as endocytosis, transfection agents, magnetofection, and electroporation);

ii. *In-vivo* cell labelling by reticuloendothelial system (RES) through systemic application; and receptor-mediated binding and internalization of SPION by targeted cells (e.g., targeted labelling and imaging). Stem cell tracking and monitoring for cell transplantation therapy reasons is one of the uses for cell labelling [52]. Living organisms, such as peptides, proteins, large molecules (such as cell receptors) or structural components of cell membranes (such as glycoproteins or cholesterol)—should make it possible to recognize activated cells, organs, or pathogenetic states for exact cell and tissue instruction to achieve therapeutic success (**Table 3**) [4].

### 3.4 Tissue engineering

The creation of functional replacement tissues or organs using patient cells has been proposed as a therapeutic strategy. These tissues or organs can develop ex vivo (in a bioreactor) for later implantation or in the patient's body at the site of the defect. It was suggested that nanomagnetic actuation be used as a mechanical simulator in TE and regenerative medicine. Bone TE bioreactors are created employing mechanical actuators made of magnetic nanoparticles. The applications of MNPs in bone tissue engineering can be expanded by the use of an external magnetic field to regulate their movement and operation. An oscillating external field in conjunction with MNPs inside a defect may increase cell induction and remotely send biomechanical cues to the cells to promote osteogenesis (**Figure 4**) [56–58].

| General name | Short name | Coating | Applications | Relaxivity | Dcc | Dhb | Clinically approved |
|---|---|---|---|---|---|---|---|
| Ferumoxide | AMI-25 | Dextran | Cell labeling, liver imaging, CNS imaging | r1 = 10.1<br>r2 = 120 | 5 | 120–180 | e |
| Feruglose | NC100150 | PEGylated starch | Blood pool agent | n.a. | n.a. | 20 | |
| Ferumoxsil | AMI-121 | Siloxane | Oral GI imaging | r1 = 2<br>r2 = 47 | 8.4 | 300 | e |
| Ferumoxtran | AMI-227 | Dextran | CNS imaging, blood pool imaging, cell labeling, macrophage imaging, lymph node imaging | r1 = 9.9<br>r2 = 65 | 5.9 | 15–30 | |
| Ferumoxytol | Code7228 | Carboxymethyl-dextran | Blood pool agent, cellular labeling, iron replacement therapy in patients with chronic kidney failure, lymph node imaging | r1 = 15<br>r2 = 89 | N.a. | 17–31 | f |
| Ferucarbotran | SHU-555A | Carboxydextran | Liver imaging, CNS imaging, cell labeling | r1 = 9.7<br>r2 = 189 | 4 | 62 | d |

**Table 3.**
*Showing the clinically approved iron oxide nanoparticles examples [42, 53–55].*

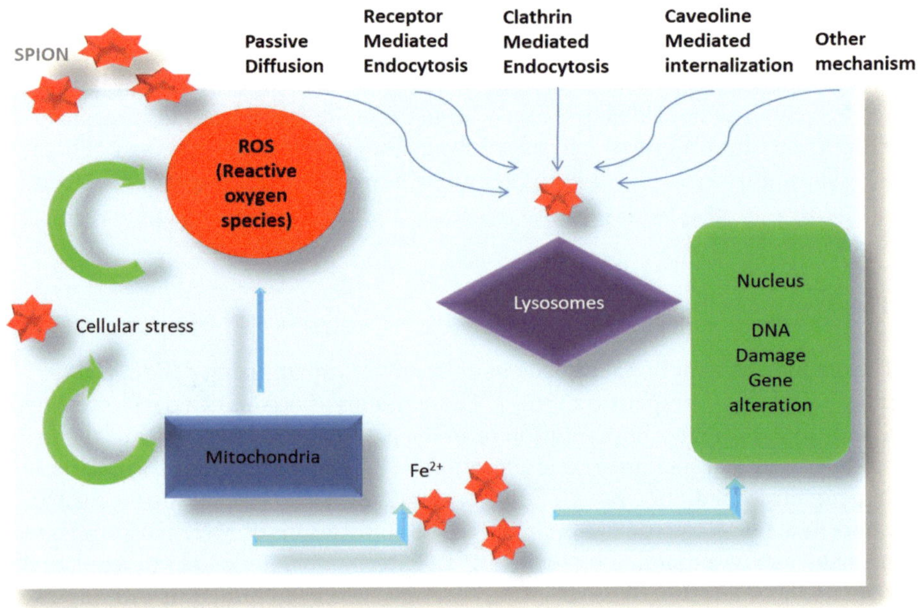


**Figure 4.**
*Showing mechanism involving cell damage persuade by SPION [56].*

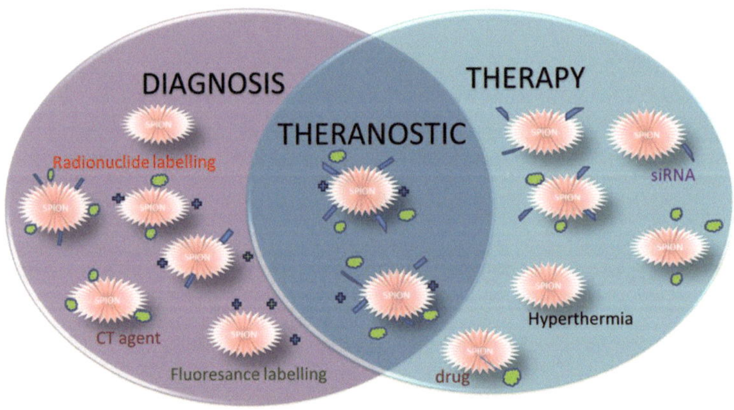


**Figure 5.**
*Showing combination of diagnosis and therapy effect of SPION [52].*

### 3.5 Magnetic particle imaging

SPION mainly used as to detect substance in the magnetic particle imaging (MPI). MPI was announce in 2005, novel pictorial representation method [59]. Generally, the SPION are superparamagnetic. After the action of the magnetic field is turned off, superparamagnetic particle will not remain magnetized. Due to Brownian and Neel relaxation, the magnetization direction can change even at ambient temperature when thermal stimulation occurs. High spatial sensitivity and sensitivity, along with the potential for high-quality real-time imaging, are the benefits of MPI above existing imaging approaches. The spatial bioavailability of the particles affects the

MPI approach's ability to produce high-quality images. Producing tracers with the best magnetically particle spectroscopy (MPS) efficiency is the difficulty of SPION synthesis (**Figure 5**) [60].

## 4. SPIONs: a better alternative for theranostic agent

The fusion of diagnostics and therapy is known as theranostics or theragnostics, and it can be divided into three types: the simultaneous use of therapeutics and diagnostics, identification followed by therapy, and therapy followed by diagnosis. Theranostics agents are created with the intent of detecting and treating disease at an early stage, monitoring the effectiveness of the therapeutic process, and minimizing time wasted during diagnosis and treatment. Theranostics nanomedicine, also known as nano-theranostics, is based on the aggregation of diagnostic and therapeutic substances into nanocarriers for use in medicine, including liposomes, micelles, carbon nanotubes, nanoparticles, and polymer-based nanomaterials. The imaging component of nanotheranostics relies on the use of fluorescent dyes like quantum dots (for optical imaging), magnetic nanoparticles, such as SPION (for MRI), radionuclides and heavy metals, such as iodine [52]. For the treatment of the diseases, theranostic medicines are used. Theranostic nanomedicine aims to reduce systemic toxicity related to cancer treatment while also improving cancer detection and treatment effectiveness (**Figure 6**) [52].

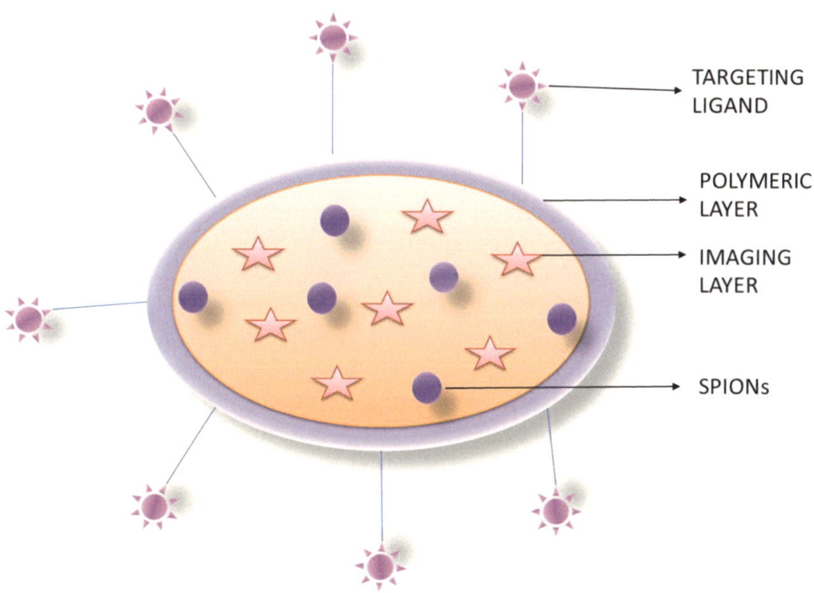


**Figure 6.**
*Shows the representation of theranostic system [61].*

## 5. Toxicity of SPION

Studies on the effects of SPION on human and animal cells at the cellular level revealed that SPION can enter cells through passive diffusion as well as endocytosis

**Figure 7.**
*Shows toxic effects of SPION [70].*

and have a number of harmful effects by changing gene expression and producing oxidative radicals [62]. Toxic or cytotoxic effects from SPION with altered physicochemical properties are possible. High levels of free ferric ions in that tissue could consequently cause an imbalance in homeostasis and abnormal cellular responses like osmotic damage, cytotoxicity, epigenetic events, DNA damage, and inflammatory disorders, which could result in cancer development or significantly affect subsequent generations [63–69]. In conclusion, toxicity of SPION, despite being suspected to be low, has not yet been adequately established because human epidemiological studies are almost nonexistent, in vivo studies are scarce, and results from in vitro studies are frequently incongruent. In the field of biomedical applications, a lot of future work has still to be done. For this, the requirement is to understood the interactions and the harmful health consequences of these nanoparticles on cellular system (**Figure** 7) [71].

## 6. Biosynthesis: an approach to overcome toxicity

One-pot biosynthetic pathway of $\gamma$-$Fe_2O_3$ NPs using natural plant extracts as reducing and capping agents has been promoted in recent years as an easy, affordable, and environmentally friendly substitute for chemical and physical methods of producing nanoparticles. The world economic scientific community is in urgent need of simple, affordable methods to produce ultrasmall (2–20 nm) maghemite nanoparticles using naturally existing plant materials. Single phase $\gamma$-$Fe_2O_3$ NPs can be synthesized by using *Ficus carica, Wedelia urticifolia, Plantago major, Pisum sativum, Citrus paradisi, Hibiscus sabdariffa, Ruellia tubercosa* as plant extract that shows antioxidant as well as catalytic activity [72, 73]. Magnetite nanoparticles has also been developed using Andean blackberry leaf extract via green synthesis approach. Environmentally friendly and appealing, this straightforward, safe, and inexpensive phytosynthesis of $Fe_3O_4$ NPs can generate substantial quantities for a variety of nanotechnology applications [74].

## 7. Conclusion

At the molecular and cellular level, SPION are an alluring platform for cell tracking, tumor diagnosis, and drug delivery owing to their unique magnetic properties and capacity to act as theranostic agents. We have described significant innovations in superparamagnetic Iron Oxide nanoparticles for biomedical applications in this book chapter. Superparamagnetic nanoparticles have been created and synthesized using a variety of chemistry principles. These distinctive structured nanoparticles, such as magneto-core-shell nanoparticles, magneto-micelles, and magnetosomes, promise applications in biomedicine for the detection of bacteria, proteins, and cells as well as benefits for contrast agents and drug delivery. Recent developments in nanotechnology and nanomaterials have led to the development of superparamagnetic nanoparticles with tunable size, morphology, and relaxivity. High $r_2$ relaxivity, appropriate particle diameters for long circulation, permeation, and immobilization of biomolecules, a narrow size distribution for a uniform response to an external magnetic field, biodegradability and stimuli responsiveness for controlled loading, and $r_2$ relaxivity are key characteristics that need to be better controlled in future design and development of effective superparamagnetic nanoparticles for biomedical applications. Additionally, establishing a strong structure-pharmacokinetics relationship is likely a crucial aspect of superparamagnetic research and requires additional research evidence. Superparamagnetic nanoparticles may soon be used for disease theranostics in translational medicine once these problems are resolved.

## Acknowledgements

Dr. Pramod Kumar would like to sincerely acknowledge the start-up funding (SERB-SRG/2020/000381) provided by the Central University of Himachal Pradesh, India.

## Conflict of interest

The authors declare no conflict of interest.

## Nomenclature

| | |
|---|---|
| SPIONs | super paramagnetic iron oxide nanoparticles. |
| MRI | magnetic resonance imaging. |

| | |
|---|---|
| Dc | core size diameter (nm) determined by laser light scattering. |
| Dh | hydrodynamic diameter (nm) determined by laser light scattering. |
| d | only limited countries available. |
| e | withdrawn from the market. |
| f | withdrawn from EU market. |
| RM | regenerative medicine. |
| TE | tissue engineering. |
| ROS | reactive oxygen species. |

## Author details

Nancy Jaswal[1], Purnima Justa[1], Hemant Kumar[2], Deepshikha[1], Krishna[1], Balaram Pani[3] and Pramod Kumar[1]*

1 Department of Chemistry and Chemical Science, School of Physical and Material Sciences, Central University of Himachal Pradesh, Dharamshala, India

2 Department of Chemistry, Ramjas College, University of Delhi, Delhi, India

3 Department of Chemistry, Bhaskaracharya College of Applied Sciences, University of Delhi, Delhi, India

*Address all correspondence to: pramodgang03@gmail.com; pramodgang03@hpcu.ac.in

IntechOpen

© 2023 The Author(s). Licensee IntechOpen. This chapter is distributed under the terms of the Creative Commons Attribution License (http://creativecommons.org/licenses/by/3.0), which permits unrestricted use, distribution, and reproduction in any medium, provided the original work is properly cited.

## References

[1] Lassenberger A, Scheberl A, Stadlbauer A, Stiglbauer A, Helbich T, Reimhult E. Individually stabilized, superparamagnetic nanoparticles with controlled shell and size leading to exceptional stealth properties and high relaxivities. ACS Applied Materials & Interfaces. 2017;**9**(4):3343-3353. DOI: 10.1021/acsami.6b12932

[2] Uthaman S, Lee SJ, Cherukula K, Cho CS, Park IK. Polysaccharide-coated magnetic nanoparticles for imaging and gene therapy. BioMed Research International. 2015;**2015**:1-14. DOI: 10.1155/2015/959175

[3] Sabale S, Kandesar P, Jadhav V, Komorek R, Motkuri RK, Yu XY. Recent developments in the synthesis, properties, and biomedical applications of core/shell superparamagnetic iron oxide nanoparticles with gold. Biomaterials Science. 2017;**5**(11):2212-2225. DOI: 10.1039/C7BM00723J

[4] Almeida AF, Vinhas A, Gonçalves AI, Miranda MS, Rodrigues MT, Gomes ME. Magnetic triggers in biomedical applications–prospects for contact free cell sensing and guidance. Journal of Materials Chemistry B. 2021;**9**(5):1259-1271. DOI: 10.1039/d0tb02474k

[5] Yoffe S, Leshuk T, Everett P, Gu F. Superparamagnetic iron oxide nanoparticles (SPIONs): Synthesis and surface modification techniques for use with MRI and other biomedical applications. Current Pharmaceutical Design. 2013;**19**(3):493-509. DOI: 10.2174/138161213804143707

[6] Amiri H, Saeidi K, Borhani P, Manafirad A, Ghavami M, Zerbi V. Alzheimer's disease: Pathophysiology and applications of magnetic nanoparticles as MRI theranostic agents. ACS Chemical Neuroscience. 2013;**4**(11):1417-1429. DOI: 10.1021/cn4001582

[7] Wang Y, Ye F, Jeong EK, Sun Y, Parker DL, Lu ZR. Noninvasive visualization of pharmacokinetics, biodistribution and tumor targeting of poly [N-(2-hydroxypropyl) methacrylamide] in mice using contrast enhanced MRI. Pharmaceutical Research. 2007;**24**(6):1208-1216. DOI: 10.1007/s11095-007-9252-1

[8] Dulińska-Litewka J, Łazarczyk A, Hałubiec P, Szafrański O, Karnas K, Karewicz A. Superparamagnetic iron oxide nanoparticles—Current and prospective medical applications. Materials. 2019;**12**(4):617. DOI: 10.3390/ma12040617

[9] Friedrich RP, Janko C, Unterweger H, Lyer S, Alexiou C. SPIONs and magnetic hybrid materials: Synthesis, toxicology and biomedical applications. Physical Sciences Reviews. 2021:1-30. DOI: 10.1515/psr-2019-0093

[10] Wu W, Wu Z, Yu T, Jiang C, Kim WS. Recent progress on magnetic iron oxide nanoparticles: Synthesis, surface functional strategies and biomedical applications. Science and Technology of Advanced Materials. 2015;**16**(2):023501. DOI: 10.1088/1468-6996/16/2/023501

[11] Laurent S, Forge D, Port M, Roch A, Robic C, Vander Elst L, et al. Magnetic iron oxide nanoparticles: Synthesis, stabilization, vectorization, physicochemical characterizations, and biological applications. Chemical Reviews. 2008;**108**(6):2064-2110. DOI: 10.1021/cr068445e

[12] Hasany SF, Ahmed I, Rajan J, Rehman A. Systematic review of the preparation techniques of iron oxide magnetic nanoparticles. Nanoscience and Nanotechnology. 2012;**2**(6):148-158. DOI: 10.5923/j.nn.20120206.01

[13] Sodipo BK, Aziz AA. Recent advances in synthesis and surface modification of superparamagnetic iron oxide nanoparticles with silica. Journal of Magnetism and Magnetic Materials. 2016;**416**:275-291. DOI: 10.1016/j.jmmm.2016.05.019

[14] Remya NS, Syama S, Sabareeswaran A, Mohanan PV. Toxicity, toxicokinetics and biodistribution of dextran stabilized iron oxide nanoparticles for biomedical applications. International Journal of Pharmaceutics. 2016;**511**(1):586-598. DOI: 10.1016/j.ijpharm.2016.06.119

[15] Petcharoen K, Sirivat AJ. Synthesis and characterization of magnetite nanoparticles via the chemical co-precipitation method. Materials Science and Engineering: B. 2012;**177**(5):421-427. DOI: 10.1016/j.mseb.2012.01.003

[16] Liu Y, Jia S, Wu Q, Ran J, Zhang W, Wu S. Studies of Fe3O4-chitosan nanoparticles prepared by co-precipitation under the magnetic field for lipase immobilization. Catalysis Communications. 2011;**12**(8):717-720. DOI: 10.1016/j.catcom.2010.12.032

[17] Wu S, Sun A, Zhai F, Wang J, Xu W, Zhang Q, et al. Fe3O4 magnetic nanoparticles synthesis from tailings by ultrasonic chemical co-precipitation. Materials Letters. 2011;**65**(12):1882-1884. DOI: 10.1016/j.matlet.2011.03.065

[18] Pereira C, Pereira AM, Fernandes C, Rocha M, Mendes R, Fernández-García MP, et al. Superparamagnetic MFe2O4 (M= Fe, Co, Mn) nanoparticles: Tuning the particle size and magnetic properties through a novel one-step coprecipitation route. Chemistry of Materials. 2012;**24**(8):1496-1504. DOI: 10.1021/cm300301c

[19] Suh SK, Yuet K, Hwang DK, Bong KW, Doyle PS, Hatton TA. Synthesis of nonspherical superparamagnetic particles: in situ coprecipitation of magnetic nanoparticles in microgels prepared by stop-flow lithography. Journal of the American Chemical Society. 2012;**134**(17):7337-7343. DOI: 10.1021/ja209245v

[20] Roy E, Patra S, Madhuri R, Sharma PK. Stimuli-responsive poly (N-isopropyl acrylamide)-co-tyrosine@ gadolinium: Iron oxide nanoparticle-based nanotheranostic for cancer diagnosis and treatment. Colloids and Surfaces B: Biointerfaces. 2016;**142**:248-258. DOI: 10.1016/j.colsurfb.2016.02.053

[21] Park J, An K, Hwang Y, Park JG, Noh HJ, Kim JY, et al. Ultra-large-scale syntheses of monodisperse nanocrystals. Nature materials. 2004;**3**(12):891-895. DOI: 10.1038/nmat1251

[22] Hufschmid R, Arami H, Ferguson RM, Gonzales M, Teeman E, Brush LN, et al. Synthesis of phase-pure and monodisperse iron oxide nanoparticles by thermal decomposition. Nanoscale. 2015;**7**(25):11142-11154. DOI: 10.1039/C5NR01651G

[23] Sharifi I, Shokrollahi H, Amiri S. Ferrite-based magnetic nanofluids used in hyperthermia applications. Journal of Magnetism and Magnetic Materials. 2012;**324**(6):903-915. DOI: 10.1016/j.jmmm.2011.10.017

[24] Lee Y, Lee J, Bae CJ, Park JG, Noh HJ, Park JH, et al. Large-scale synthesis

of uniform and crystalline magnetite nanoparticles using reverse micelles as nanoreactors under reflux conditions. Advanced Functional Materials. 2005;**15**(3):503-509. DOI: 10.1002/adfm.200400187

[25] Tartaj P, Serna CJ. Microemulsion-assisted synthesis of tunable superparamagnetic composites. Chemistry of Materials. 2002;**14**(10):4396-4402. DOI: 10.1021/cm021214d

[26] Puscasu E, Sacarescu L, Lupu N, Grigoras M, Oanca G, Balasoiu M, et al. Iron oxide-silica nanocomposites yielded by chemical route and sol–gel method. Journal of Sol-Gel Science and Technology. 2016;**79**(3):457-465. DOI: 10.1007/s10971-016-3996-1

[27] Fernandes MT, Garcia RB, Leite CA, Kawachi EY. The competing effect of ammonia in the synthesis of iron oxide/silica nanoparticles in microemulsion/sol–gel system. Colloids and Surfaces A: Physicochemical and Engineering Aspects. 2013;**422**:136-142. DOI: 10.1016/j.colsurfa.2013.01.025

[28] Sun S, Zeng H, Robinson DB, Raoux S, Rice PM, Wang SX, et al. Monodisperse mfe2o4 (m= fe, co, mn) nanoparticles. Journal of the American Chemical Society. 2004;**126**(1):273-279. DOI: 10.1021/ja0380852

[29] Lee YT, Woo K, Choi KS. Preparation of water-dispersible and biocompatible iron oxide nanoparticles for MRI agent. IEEE Transactions on Nanotechnology. 2008;7(2):111-114. DOI: 10.1109/TNANO.2007.909949

[30] Mahmoudi M, Sant S, Wang B, Laurent S, Sen T. Superparamagnetic iron oxide nanoparticles (SPIONs): Development, surface modification and applications in chemotherapy. Advanced Drug Delivery Reviews. 2011;**63**(1-2):24-46. DOI: 10.1016/j.addr.2010.05.006

[31] Sharapova VA, Uimin MA, Mysik AA, Ermakov AE. Heat release in magnetic nanoparticles in AC magnetic fields. The Physics of Metals and Metallography. 2010;**110**(1):5-12. DOI: 10.1134/S0031918X10070021

[32] Buyukhatipoglu K, Morss CA. Controlled flame synthesis of αFe2O3 and Fe3O4 nanoparticles: Effect of flame configuration, flame temperature, and additive loading. Journal of Nanoparticle Research. 2010;**12**(4):1495-1508. DOI: 10.1007/s11051-009-9724-9

[33] Kang YS, Risbud S, Rabolt JF, Stroeve P. Synthesis and characterization of nanometer-size Fe3O4 and γ-Fe2O3 particles. Chemistry of Materials. 1996;**8**(9):2209-2211. DOI: 10.1021/cm960157j

[34] Qu S, Yang H, Ren D, Kan S, Zou G, Li D, et al. Magnetite nanoparticles prepared by precipitation from partially reduced ferric chloride aqueous solutions. Journal of Colloid and Interface Science. 1999;**215**(1):190-192. DOI: 10.1006/jcis.1999.6185

[35] Chen D, Xu R. Hydrothermal synthesis and characterization of nanocrystalline Fe3O4 powders. Materials Research Bulletin. 1998;**33**(7):1015-1021. DOI: 10.1016/S0025-5408(98)00073-7

[36] Arbain R, Othman M, Palaniandy S. Preparation of iron oxide nanoparticles by mechanical milling. Minerals Engineering. 2011;**24**(1):1-9. DOI: 10.1016/j.mineng.2010.08.025

[37] Janot R, Guérard D. One-step synthesis of maghemite nanometric powders by ball-milling. Journal

of Alloys and Compounds. 2002;333(1-2):302-7DOI: 10.1016/S0925-8388(01)01737-6

[38] Islam MN, Jeong JR, Kim C. A facile route to sonochemical synthesis of magnetic iron oxide (Fe3O4) nanoparticles. Thin Solid Films. 2011;**519**(23):8277-8279. DOI: 10.1016/j.tsf.2011.03.108

[39] Mirza S, Ahmad MS, Shah MI, Ateeq M. Magnetic nanoparticles: Drug delivery and bioimaging applications. In: Metal Nanoparticles for Drug Delivery and Diagnostic Applications. Elsevier; 2020:189-213. DOI: 10.1016/B978-0-12-816960-5.00011-2

[40] Pinkas J, Reichlova V, Zboril R, Moravec Z, Bezdicka P, Matejkova J. Sonochemical synthesis of amorphous nanoscopic iron (III) oxide from Fe (acac) 3. Ultrasonics Sonochemistry. 2008;**15**(3):257-264. DOI: 10.1016/j.ultsonch.2007.03.009

[41] Zhang S, Zhang Y, Wang Y, Liu S, Deng Y. Sonochemical formation of iron oxide nanoparticles in ionic liquids for magnetic liquid marble. Physical Chemistry Chemical Physics. 2012;**14**(15):5132-5138. DOI: 10.1039/C2CP23675C

[42] Dadfar SM, Roemhild K, Drude NI, von Stillfried S, Knüchel R, Kiessling F, et al. Iron oxide nanoparticles: Diagnostic, therapeutic and theranostic applications. Advanced Drug Delivery Reviews. 2019;**138**:302-325. DOI: 10.1016/j.addr.2019.01.005

[43] Darmawan A, Smart S, Julbe A, Diniz da Costa JC. Iron oxide silica derived from sol-gel synthesis. Materials. 2011;**4**(2):448-456. DOI: 10.3390/ma4020448

[44] Kohler N, Sun C, Fichtenholtz A, Gunn J, Fang C, Zhang M. Methotrexate-immobilized poly (ethylene glycol) magnetic nanoparticles for MR imaging and drug delivery. Small. 2006;**2**(6):785-792. DOI: 10.1002/smll.200600009

[45] Kohler N, Sun C, Wang J, Zhang M. Methotrexate-modified superparamagnetic nanoparticles and their intracellular uptake into human cancer cells. Langmuir. 2005;**21**(19):8858-8864. DOI: 10.1021/la0503451

[46] Guo M, Que C, Wang C, Liu X, Yan H, Liu K. Multifunctional superparamagnetic nanocarriers with folate-mediated and pH-responsive targeting properties for anticancer drug delivery. Biomaterials. 2011;**32**(1):185-194. DOI: 10.1016/j.biomaterials.2010.09.077

[47] Lin JJ, Chen JS, Huang SJ, Ko JH, Wang YM, Chen TL, et al. Folic acid–Pluronic F127 magnetic nanoparticle clusters for combined targeting, diagnosis, and therapy applications. Biomaterials. 2009;**30**(28):5114-5124. DOI: 10.1016/j.biomaterials.2009.06.004

[48] Kim DK, Chang JH, Kang YJ. Efficient internalization of peptide-conjugated SPIONs in dendritic cells for tumor targeting. Journal of Nanoscience and Nanotechnology. 2012;**12**(7):5191-5198. DOI: 10.1166/jnn.2012.6379

[49] Cho HS, Dong Z, Pauletti GM, Zhang J, Xu H, Gu H, et al. Fluorescent, superparamagnetic nanospheres for drug storage, targeting, and imaging: A multifunctional nanocarrier system for cancer diagnosis and treatment. ACS Nano. 2010;**4**(9):5398-5404. DOI: 10.1021/nn101000e

[50] Taratula O, Garbuzenko O, Savla R, Andrew Wang Y, He H, Minko T. Multifunctional nanomedicine platform

for cancer specific delivery of siRNA by superparamagnetic iron oxide nanoparticles-dendrimer complexes. Current Drug Delivery. 2011;**8**(1):59-69. DOI: 10.2174/156720111793663642

[51] Leuschner C, Kumar CS, Hansel W, Hormes J. Targeting breast cancer cells and their metastases through luteinizing hormone releasing hormone (LHRH) receptors using magnetic nanoparticles. Journal of Biomedical Nanotechnology. 2005;**1**(2):229-233. DOI: 10.1166/jbn.2005.027

[52] Zarepour A, Zarrabi A, Khosravi A. Spions as nano-theranostics agents. In: SPIONs as Nano-Theranostics Agents. Singapore: Springer; 2017. pp. 1-44. DOI: 10.1007/978-981-10-3563-0_1

[53] Iv M, Telischak N, Feng D, Holdsworth SJ, Yeom KW, Daldrup-Link HE. Clinical applications of iron oxide nanoparticles for magnetic resonance imaging of brain tumors. Nanomedicine. 2015;**10**(6):993-1018. DOI: 10.2217/nnm.14.203

[54] Qiao R, Yang C, Gao M. Superparamagnetic iron oxide nanoparticles: From preparations to in vivo MRI applications. Journal of Materials Chemistry. 2009;**19**(35):6274-6293. DOI: 10.1039/B902394A

[55] Jin R, Lin B, Li D, Ai H. Superparamagnetic iron oxide nanoparticles for MR imaging and therapy: Design considerations and clinical applications. Current Opinion in Pharmacology. 2014;**18**:18-27. DOI: 10.1016/j.coph.2014.08.002

[56] Nabavinia M, Beltran-Huarac J. Recent progress in iron oxide nanoparticles as therapeutic magnetic agents for cancer treatment and tissue engineering. ACS Applied Bio Materials. 2020;**3**(12):8172-8187. DOI: 10.1021/acsabm.0c00947

[57] Dasari A, Xue J, Deb S. Magnetic nanoparticles in bone tissue engineering. Nanomaterials. 2022;**12**(5):757. DOI: 10.3390/nano12050757

[58] Gobbo OL, Sjaastad K, Radomski MW, Volkov Y, Prina-Mello A. Magnetic nanoparticles in cancer theranostics. Theranostics. 2015;**5**(11):1249. DOI: 10.7150/thno.11544

[59] Lindemann A, Lüdtke-Buzug K, Fräderich BM, Gräfe K, Pries R, Wollenberg B. Biological impact of superparamagnetic iron oxide nanoparticles for magnetic particle imaging of head and neck cancer cells. International Journal of Nanomedicine. 2014;**9**:5025. DOI: 10.2147/IJN.S63873

[60] Zanganeh S, Aieneravaie M, Erfanzadeh M, Ho JQ, Spitler R. Magnetic particle imaging (MPI). In: Iron Oxide Nanoparticles for Biomedical Applications. Elsevier; 2018:115-133. DOI: 10.1016/B978-0-08-101925-2.00004-8

[61] Bai RG, Muthoosamy K, Manickam S. Nanomedicine in theranostics. In: Nanotechnology Applications for Tissue Engineering. William Andrew Publishing; 2015:195-213. DOI: 10.1016/B978-0-323-32889-0.00012-1

[62] Barhoumi L, Dewez D. Toxicity of superparamagnetic iron oxide nanoparticles on green alga Chlorella vulgaris. BioMed Research International. 2013;**2013**:1-11. DOI: 10.1155/2013/647974

[63] Vakili-Ghartavol R, Momtazi-Borojeni AA, Vakili-Ghartavol Z, Aiyelabegan HT, Jaafari MR,

Rezayat SM, et al. Toxicity assessment of superparamagnetic iron oxide nanoparticles in different tissues. Artificial Cells, Nanomedicine, and Biotechnology. 2020;**48**(1):443-451. DOI: 10.1080/21691401.2019.1709855

[64] Kumar H, Agnihotri S, Roy I, Pani B, Kumar P. Microemulsion mediated multifunction of doxorubicin encapsulated Core–Shell iron oxide/ Ormosil nanoparticles as efficient magnetically-guided delivery, bioimaging and In-vitro studies. Advanced Science, Engineering and Medicine. 2020;**12**(9):1166-1173. DOI: 10.1166/asem.2020.2674

[65] Kumar P, Agnihotri S, Roy I. Preparation and characterization of superparamagnetic iron oxide nanoparticles for magnetically guided drug delivery. International Journal of Nanomedicine. 2018;**13**(T-NANO 2014 Abstracts:43. DOI: 10.2147/IJN.S125002

[66] Kumar P, Agnihotri S, Roy I. Synthesis of photoactive SPIONs doped with visible light activated photosensitizer. Journal of Nanomedicine Nanotechnology. 2016;**7**(392):2. DOI: 10.4172/2157-7439.1000392

[67] Kumar P, Agnihotri S, Roy I. Synthesis of dox drug conjugation and citric acid stabilized superparamagnetic iron-oxide nanoparticles for drug delivery. Biochemistry & Physiology. 2016;**5**(194):2. DOI: 10.4172/ 2168-9652.1000194

[68] Kumar H, Kumar J, Pani B, Kumar P. Multifunctional folic acid-coated and doxorubicin encapsulated mesoporous silica nanocomposites (FA/DOX@ silica) for cancer therapeutics, bioimaging and invitro studies. Chemistry Select. 2022;**7**(44):e202203113. DOI: 10.1002/ slct.202203113

[69] Kumar H, Kumar J, Pani B, Kumar P. In-vitro and bioimaging studies of mesoporous silica nanocomposites encapsulated iron-oxide and loaded doxorubicin drug (DOX/IO@silica) as magnetically guided drug delivery system. Current Pharmaceutical Biotechnology. 2022. DOI: 10.2174/13892 01023666220428084920

[70] Singh N, Jenkins GJ, Asadi R, Doak SH. Potential toxicity of superparamagnetic iron oxide nanoparticles (SPION). Nano Reviews. 2010;**1**(1):5358. DOI: 10.3402/nano. v1i0.5358

[71] Valdiglesias V, Fernández-Bertólez N, Kiliç G, Costa C, Costa S, Fraga S, et al. Are iron oxide nanoparticles safe? Current knowledge and future perspectives. Journal of Trace Elements in Medicine and Biology. 2016;**38**:53-63. DOI: 10.1016/j.jtemb.2016.03.017

[72] Kumar B, Smita K, Galeas S, Guerrero VH, Debut A, Cumbal L. One-pot biosynthesis of maghemite (γ-Fe 2 O 3) nanoparticles in aqueous extract of ficus carica fruit and their application for antioxidant and 4-nitrophenol reduction. Waste and Biomass Valorization. 2021;**12**:3575-3587. DOI: 10.1007/s12649-020-01279-9

[73] Kumar B. Green synthesis of gold, silver, and iron nanoparticles for the degradation of organic pollutants in wastewater. Journal of Composites Science. 2021;**5**(8):219. DOI: 10.3390/ jcs5080219

[74] Kumar B, Smita K, Cumbal L, Debut A, Galeas S, Guerrero VH. Phytosynthesis and photocatalytic activity of magnetite (Fe3O4) nanoparticles using the Andean blackberry leaf. Materials Chemistry and Physics. 2016;**179**:310-315. DOI: 10.1016/j.matchemphys.2016.05.045

Chapter 4

# The Characteristics and Reduction of Wustite

*Zuoliang Zhang*

## Abstract

The wustite is the most difficult to be reduced in the three iron oxides of hematite, magnetite, and wustite. To introduce the structure, property, and characteristic of the wustite completely is necessary. CO and C are the common reducing agents. However, C can reduce the wustite to iron more easily than CO in high temperature. Especially, reducing carbon dioxide emissions is so important that hydrogen is the alternative to carbon. The thermodynamics and kinetics of the three reductants to reduce the wustite will be proposed in detail. When the temperature is lower than 810°C (1083 K), the reduction ability of CO is stronger than that of $H_2$. If it is higher than 810°C (1083 K), the opposite is true. In the meantime, the difference among the three reductants to reduce the wustite can be understood clearly. There are some other aspects that will be introduced. When the temperature is 1173 K, according to calculation, considering the theoretical amount and the chemical balance of the reduction reaction, the amount of hydrogen required to be put into the reactor is 952 $Nm^3$/t Fe. Wustite will be reduced by pure hydrogen in some reactors in the foreseeable future.

**Keywords:** wustite, reduction, hydrogen, iron oxides, reduction kinetics

## 1. Introduction

Pure wustite is not found in nature because it is not stable in thermodynamics. It is the intermediate product in the reduction process of iron ore, and it is also the most difficult to reduce. It is necessary to understand the reduction of wustite by reducing agents in thermodynamics and kinetics. The reduction process of wustite in blast furnace is so important to be discussed. The use of pure hydrogen in ironmaking in the future is desirable.

## 2. Characteristics of wustite

The actual chemical formula of the wustite is $Fe_xO$ or $Fe_{1-y}O$, where y represents the relative number of the ferrous iron ion vacancies; x is equal to 1_y. It is cubic crystal with sodium chloride (**Figure 1**). In other words, there is no compound FeO with a theoretical oxygen content of 22.27 wt% and an atomic ratio of Fe to O of 1:1. The oxygen content varies at different temperatures, and the maximum variation

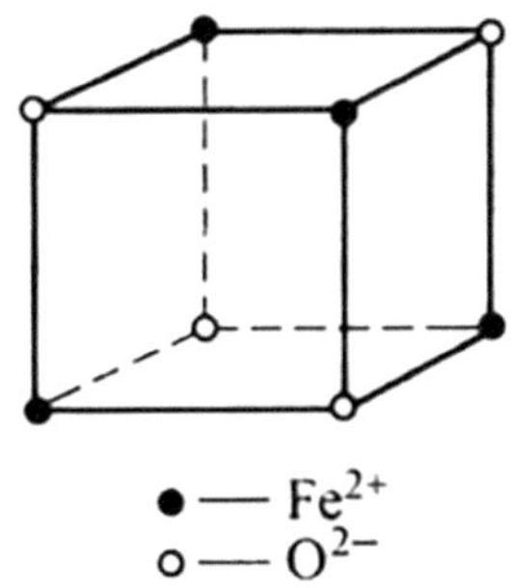


**Figure 1.**
*Cell structure of sodium chloride monomer in cubic system of FeO [1].*

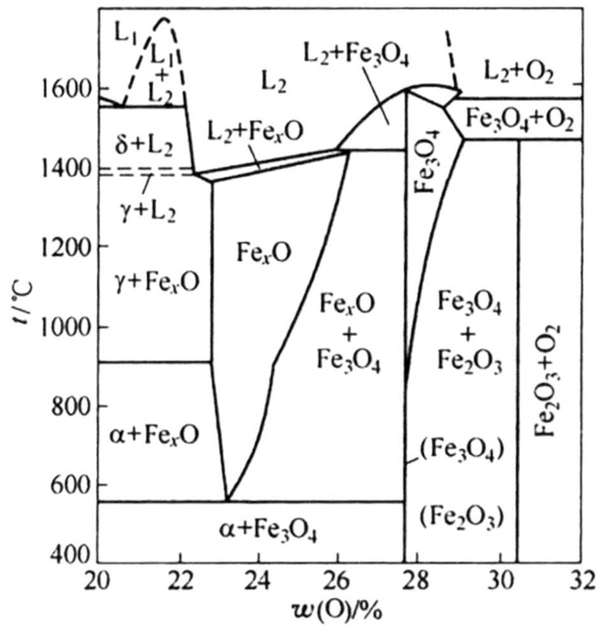


**Figure 2.**
*Phase diagram of Fe-O [1]. $L_1$-Liquid iron; $L_2$-Liquid oxide.*

range is from 23.16 wt% to 25.60 wt%. This can be understood by the phase diagram of Fe-O (**Figure 2**).

The wustite should not be stable at low temperatures. When the temperature is lower than 570°C (843 K), $Fe_xO$ will be decomposed into $Fe_3O_4$ (Magnetite) and $\alpha$Fe. For convenience, the wustite can be written as FeO to participate in chemical reactions, which is the compound with a fixed component.

## 3. Reduction thermodynamics

Here, reduction refers to capturing the oxygen combined with metal elements in the ore, which is the basic task to be completed in the smelting process. It uses a substance (reducing agent) with a stronger ability to combine with oxygen to break

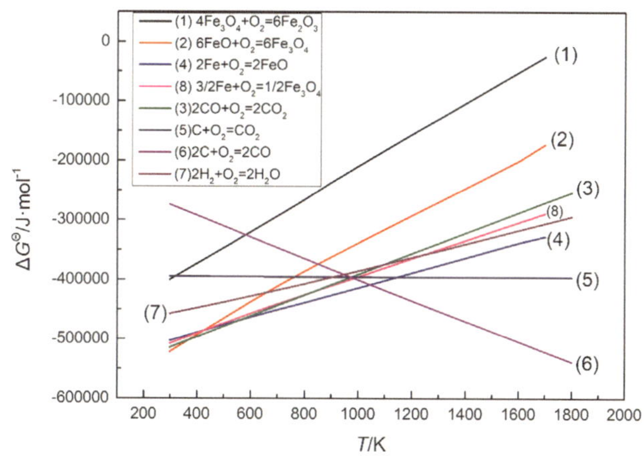


**Figure 3.**
*Ellingham diagram.*

the chemical key of metal ions and oxygen ions in ore and release metal elements. The reducing agents, such as carbon (C), carbon monoxide (CO), and hydrogen ($H_2$), are selected for industrial production.

Both production practice and scientific research have proved that no matter what reductant is used to reduce iron oxide, it changes from high-valence oxide to low-valence oxide step by step. In the laboratory, the hematite ball in the reduction process is rapidly placed in a neutral or inert atmosphere for cooling, and then its section is taken for observation. A distinct layered structure can be found. The core of the ball is unreacted, with a layer of $Fe_3O_4$ on the outside, a thin layer of wustite on the outside, and metal iron gradually thickened with the reaction on the outside. The semi-reduced ore samples obtained from the furnace during the dissection of the blast furnace also have the same shell structure.

To judge the degree of difficulty of various iron oxides being reduced by different reductants at different temperatures, the most basic basis is the graph of changes in the standard free energy of formation of various oxides with temperature, or the "oxygen potential graph." The diagram of standard free energy of formation of iron oxide and oxide generated by relevant reductant is only involved in iron ore reduction, which is called the Ellingham diagram (**Figure 3**).

According to the Ellingham diagram, the most stable of the three iron oxides is FeO. Simultaneously, FeO cannot be reduced by CO or $H_2$ above 570°C (843 K).

### 3.1 Indirect reduction

Indirect reduction means that the reductant is gaseous CO or $H_2$, and the product is $CO_2$ or $H_2O$.

$$\mathrm{FeO} + \mathrm{CO} = \mathrm{Fe} + \mathrm{CO2},\ \Delta H^{\theta}_{298} = -13190, \mathrm{J}\cdot\mathrm{mol}^{-1} \tag{1}$$

$$\lg K = \lg\frac{p_{CO_2}}{p_{CO}} = \frac{688}{T} - 0.9 \tag{2}$$

$$\mathrm{FeO} + \mathrm{H2} = \mathrm{Fe} + \mathrm{H2O},\ \Delta H^{\theta}_{298} = 28010, \mathrm{J}\cdot\mathrm{mol}^{-1} \tag{3}$$

$$\lg K = \lg\frac{p_{H_2O}}{p_{H_2}} = -\frac{1225}{T} + 0.845 \tag{4}$$

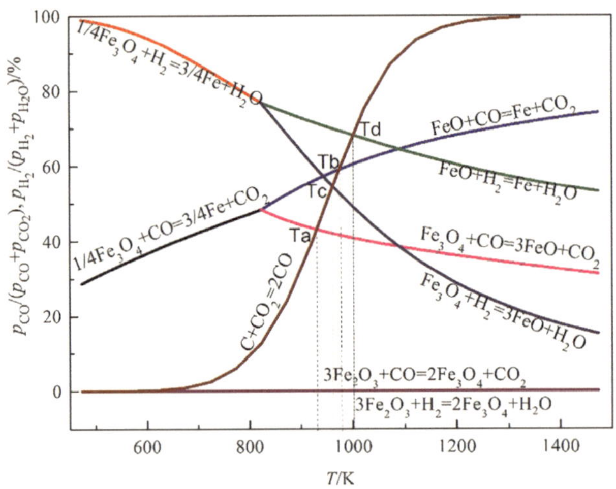


**Figure 4.**
*Equilibrium iron oxide diagram reduced by CO, C, and $H_2$.*

The reactions listed have a common feature, that is, there is a gas phase before and after the reaction and its molecular number (i.e., volume) is unchanged. Under the condition that other substances participating in the reaction are pure solid, the equilibrium state of the reaction is not affected by the total pressure of the system. The equilibrium constant of the reaction can be expressed by $K = p(CO_2)/p(CO)$ (or $p(H_2O)/p(H_2)$). Since it is independent of the total pressure, it can also be expressed as $K = \varphi(CO_2)/\varphi(CO)$ (or $\varphi(H_2O)/\varphi(H_2)$. Without considering other inert components (such as $N_2$) in the gas phase, $\varphi(CO_2) + \varphi(CO) = 100\%$ (or $\varphi(H_2O) + \varphi(H_2) = 100\%$), then the equilibrium constant or equilibrium state can also be simplified to be expressed by single value gas composition, such as $\varphi(CO_2)$ or $\varphi(CO)$ or ($\varphi(H_2O)$ or $\varphi(H_2)$). The equilibrium diagram of iron oxide reduced by CO, C, and $H_2$ is shown in **Figure 4**.

From **Figure 4**, with the promotion of the reduction reaction, the high-valence oxides with high oxygen content are reduced into low-valence oxides with low oxygen content. The reduction reaction becomes more difficult, which is reflected in the requirements of $\varphi(CO)$ in the equilibrium gas phase composition is getting higher and higher. The process from FeO to Fe is the key step.

It also can be seen from **Figure 4** that the slope of the curve is related to the thermal effect of the reaction. The exothermic reaction increases with temperature, and $\varphi(CO)$ in the equilibrium gas phase composition also increases, and the curve is inclined upward. The endothermic reaction is converse. Reduction of iron oxide by hydrogen is an endothermic reaction, and when the reducing gas is pure hydrogen, the heat carried by hydrogen is the heat source of the reduction reaction. Therefore, heating hydrogen to the target temperature is an important factor to ensure the reduction of iron oxide.

CO and $H_2$ have the same reduction capacity at 810°C (1083 K). When the temperature is lower than 810°C (1083 K), the reduction ability of CO is stronger than that of $H_2$. If it is higher than 810°C (1083 K), the opposite is true. This can be explained by the equilibrium relation of water gas displacement reaction in the C—O—H system. The reaction formula is:

$$CO + H2O = CO2 + H2, \ \Delta H^{\theta}_{298} = -41325, J \cdot mol^{-1} \quad (5)$$

$$\lg K = \lg \frac{p_{H_2} p_{CO_2}}{p_{CO} p_{H_2O}} = \frac{1591}{T} - 1.469 \tag{6}$$

For blast furnace, the existence of water gas displacement reaction makes $H_2$ have the effect of promoting CO reduction, which is equivalent to the catalyst of CO reduction reaction.

When the water gas displacement reaction is above 800°C, the reaction speed is quite fast, that is, once CO reduces from FeO to $CO_2$, due to the existence of $H_2$, it can be rapidly regenerated into CO to participate in the reduction.

Since this displacement reaction is a homogeneous reaction (reactants are all in the gas phase), at 750–800 K (477–527°C) in the blast furnace, under the catalysis of Fe, Cr, Ni, Mn, and other oxides, CO and $H_2O$ start to react, while at 1330 K (1057°C), the reaction can be carried out at a high speed without any catalyst, which can be considered as reaching an equilibrium state in the blast furnace.

In the high-temperature zone where the reduction capacity of $H_2$ is stronger than that of CO, the superposition result of $H_2$ reduction and water gas replacement is equivalent to the consumption of CO and the reduction of FeO, which improves the CO use efficiency. In the middle- and low-temperature regions where the reduction capacity of CO is stronger than that of $H_2$, although $H_2$ is superior to CO in kinetics, the amount of $H_2$ involved in indirect reduction is not more than that of CO due to the restriction of thermodynamics. Therefore, the utilization ratio of $H_2$ and CO in the blast furnace not only increases but also restricts each other.

We have already known that wustite is an oxide with an unstable oxygen content, and its reduction reaction equations are expressed as follows:

$$(O)_{FeO} + CO = CO_2 \tag{7}$$

$$(O)_{FeO} + H_2 = H_2O \tag{8}$$

The equilibrium constants of the above reactions are expressed as follows:

$$K = \frac{p_{CO_2}}{p_{CO}} \cdot \frac{1}{a_{(O)}} \tag{9}$$

$$K = \frac{p_{H_2O}}{p_{H_2}} \cdot \frac{1}{a_{(O)}} \tag{10}$$

It can be seen from the formulas that, for the wustite, when it is reduced by gaseous reductant, the equilibrium gas phase composition is not only a function of temperature but also a function of the oxygen content of the wustite.

The equilibrium diagrams of Fe—O—CO and Fe—O—$H_2$ are shown in **Figures 5** and **6**.

### 3.2 Direct reduction

Direct reduction means that the reducing agent is solid C and the product is CO.

$$FeO + C = Fe + CO, \ \Delta H^{\theta}_{298} = 152200, J \cdot mol^{-1} \tag{11}$$

$$\lg K = \lg p_{CO} = -\frac{8270}{T} + 7.89 \tag{12}$$

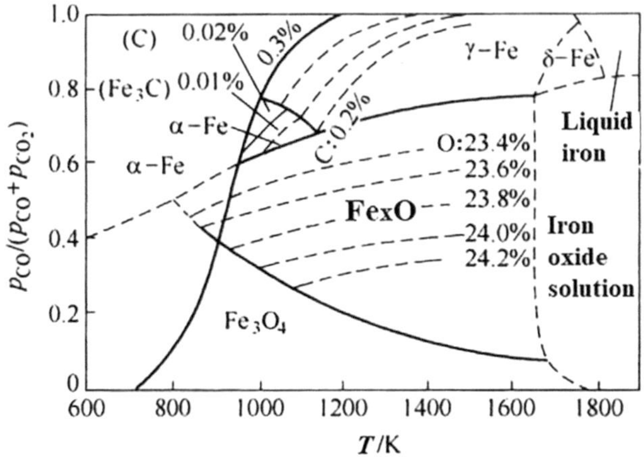


**Figure 5.**
*Equilibrium diagram of Fe—O—CO [1].*

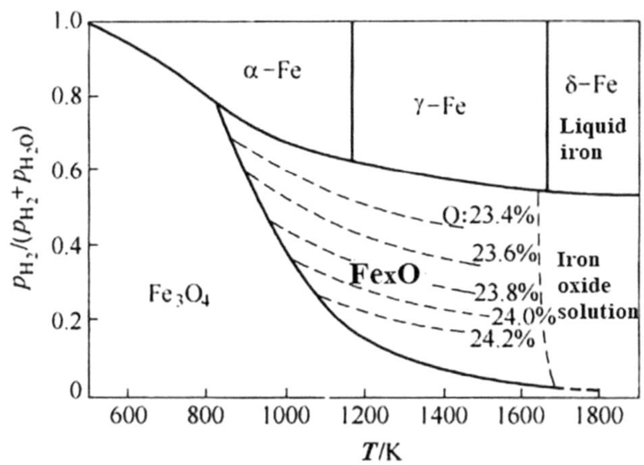


**Figure 6.**
*Equilibrium diagram of Fe—O—$H_2$ [1].*

In a blast furnace, the direct reduction does not mean that only the direct contact reaction between solid C and solid FeO can occur (except the reaction between FeO in liquid slag and solid C). On the contrary, due to the extremely poor contact conditions between the two solid phases, it is not enough to maintain the perceptible reaction rate. The actual direct reduction reaction is realized by the superposition of carbon dissolution loss reaction ($C + CO_2 = 2CO$, **Figure 4**) and water gas reaction ($H_2O + C{=}CO + H_2$) and indirect reduction reaction.

Direct reduction mainly refers to the direct consumption of solid carbon. Another characteristic of this reaction is that it is strongly endothermic, and the thermal effect is up to 2717 kJ/kg.

Since this reaction only involves one gas phase product, its equilibrium constant can be expressed by the partial pressure of CO ($p_{CO}$), and excess carbon will not be consumed due to the requirement of special equilibrium gas phase composition.

## 4. Reduction kinetics

The research on the reaction mechanism and quantitative law of the rate of gas reduction of solid iron ore has gone through a state of gradual perfection and approaching to reality from the shallow to the deep. In the late 1960s, most metallurgical workers tended to believe that the whole reduction process was composed of a series of interconnected subprocesses, and was often controlled not by a single subprocess but by two or more sub-processes. At different stages of the reduction process, the process control links may also be transformed. Due to different understandings of the role played by each process, a variety of reduction process mechanism models are proposed, such as unreacted core model, layered model, quasi-homogeneous model, and intermediate model.

Because it is of no practical significance to study the reduction kinetics of pure FeO, and the reduction of wustite to metal Fe is the most difficult and the most oxygen removal stage in the reduction process of single ore particles, the deoxidation process of $Fe_2O_3$ (hematite) and $Fe_3O_4$ is ignored, and it is believed that there is only one reaction interface. The following is a brief introduction to the "unreacted core model".

According to the sequence of iron oxide reduction, the fact that the cross section of the single ore particles in the reduction process is layered, and the unreacted core gradually shrinks with the reaction process, the mathematical expression of the overall reduction rate is derived from the basic chemical thermodynamics and dynamics principles, taking into account all the processes of the whole reduction process.

Take the cross section of ore ball with layered structure in the reduction process (**Figure** 7), and assume that the secondary process of each link is as follows.

Reduction gas diffuses outside the boundary layer → Reduction gas diffuses inside the product layer → Reduction gas adsorbs at the reaction interface → Interface chemical reaction → Oxidation gas desorbs → Oxidation gas diffuses inside the product layer → Oxidation gas diffuses outside the boundary layer.

In order to simplify the mathematical treatment, (1) the ore particles are assumed to be regular spherical with radius of $r_0$, and there is no shrinkage or expansion in the reduction process, that is, $r_0$ remains unchanged; (2) the radius of the unreacted nucleus is $r_i$, and with the progress of the reduction process, $r_i$ changes from $r_0$ at the beginning to 0 at the time of complete reduction; (3) since the reduction of Wustite to metal Fe is the most difficult stage for oxygen removal, the process of deoxidation of

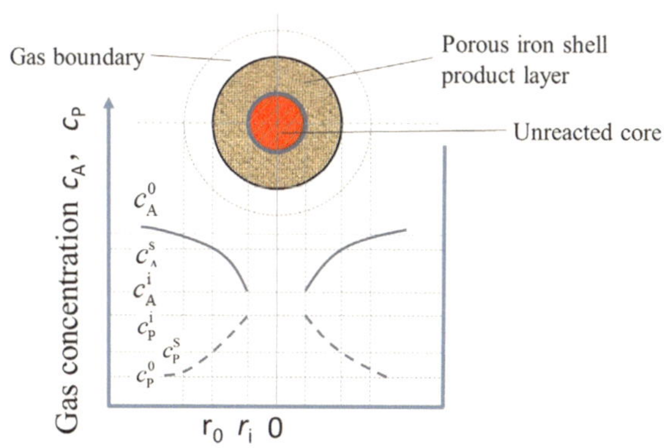


**Figure 7.**
*Unreacted core model.*

$Fe_2O_3$ and $Fe_3O_4$ is ignored, and only one reaction interface exists; (4) there is a boundary layer of adsorbed and relatively static reducing gas outside the ore ball (dotted line in **Figure** 7).

Finally, the restore process is simplified into the following five sub-processes:

1. The molecules of the gas-reducing agent pass through the gas boundary layer (external diffusion) from the gas mainstream (gas concentration: $c_A^0$) to the outer surface of the ball, and the concentration drops to $c_A^s$.

2. The gas-reducing agent molecules diffuse (can be called internal diffusion) through the porous reduction product (iron shell) layer to reach the outer surface of the unreacted core (i.e., the reaction interface), and the concentration further decreases to $c_A^i$.

3. The chemical reaction takes place on the interface, and the details, such as the adsorption of reducing gas molecules and the desorption of oxidation gas generated after the reaction, are ignored here.

4. After desorption of oxidized gas molecules, the concentration on the surface of unreacted nucleus is $c_P^i$, and it diffuses outward (also called internal diffusion) through the product layer of metal iron. When it reaches the surface of ore ball, the concentration drops to $c_P^s$.

5. The oxidation gas diffuses into the gas mainstream through the gas boundary layer, and the concentration decreases to $c_P^0$.

The change process of $c_A$ and $c_P$ along the transmission route is shown in the curve in **Figure** 7. Since the radius of the unreacted nucleus and the corresponding reaction interface area are gradually reduced, while the reaction product layer is gradually thickened at the same time, the internal diffusion path is gradually increased and the tortuosity is increased, so the internal diffusion process is not in a stable state, that is, the gas concentration gradient is not a constant independent of time. However, compared with the velocity of gas diffusion, the velocity of unreacted nuclear interface propulsion is several orders of magnitude lower. In order to simplify the mathematical process, the diffusion process can be approximately regarded as being in a stable state.

The specific derivation process of the reduction rate mathematical model is ignored here, and only the formulas of the reaction rate are given as follows:

$$R_A = \frac{c_A^0 - c_A^*}{\frac{1}{4\pi r_0^2 \cdot \beta_A} + \frac{1}{4\pi \cdot D_{eff}\left(\frac{r_0 \cdot r_i}{r_0 - r_i}\right)} + \frac{K}{\frac{1}{4\pi r_i^2} \cdot k_+(1+K)}} \tag{13}$$

$$\frac{df}{dt} = \frac{3}{r_O \rho_O} \cdot \frac{c_A^0 - c_A^*}{\frac{1}{\beta_A} + \frac{r_O}{D_{eff}}\left[(1\text{-}f)^{-1/3} - 1\right] + \frac{K}{k_+(1+K)}(1\text{-}f)^{-2/3}} \tag{14}$$

The general formula (13) of reaction rate can only be used for theoretical analysis of the parameters that affect the reduction rate. After transformation, the formula (14) expressed in measurable parameters can be used to test the experimental data to determine which process (or multiple processes) of an ore under different conditions becomes the controlling link of the reduction reaction.

## 5. Transformation of wustite in blast furnace

Once ore is added to the blast furnace, it will undergo a series of physical and chemical changes. The total residence time of ore in the blast furnace fluctuates between 5 and 8 hours. The gas-solid phase reduction process from high-valence oxide to wustite is completed in 1–2 h, and then half or more of the wustite is reduced to metallic iron in an indirect reduction way in 1–2 h. After entering the high-temperature zone higher than 1000°C (1273 K), only direct reduction can be carried out. Slag is formed after the furnace charge is heated to softening and melting temperature, and there is still a considerable amount of liquid wustite that needs to be reduced at a very fast speed by solid carbon or carbon dissolved in iron [1]. A more specific description of the FeO change process is described below [2].

The reduction process starts at temperatures of about 500°C (773 K) in the atmosphere of a reducing gas, that is, the blast furnace top gas. The reduction of hematite to magnetite takes place rather easily. The magnetite is further reduced by gas (CO and $H_2$) to wustite. All reduction is taking place by means of gas reduction ($Fe_2O_3 + CO \rightarrow 2FeO + CO_2$), and in this temperature range also for a small part by hydrogen reduction ($Fe_2O_3 + H_2 \rightarrow 2FeO + H_2O$). The gas reduction continues to a gas temperature above 1000°C (1273 K) and a reduction of iron oxide to a level of $FeO_{0.5}$. The higher the temperature, the more $H_2$ contributes to the gas reduction. The gas reduction continues to rise until the temperature has risen to that where the coke reactivity reaction begins.

If the material starts to soften and melt (around 1100°C), the direct reduction reaction ($FeO + C \rightarrow Fe + CO$) will take place. At that moment, the atomic ratio of O/Fe is slightly below 0.5 atom oxygen per atom of iron. In the case of natural gas injection, it can decrease to around 0.35 atom oxygen per atom of iron.

Softening and melting start at local chemical compositions with the lowest melting temperatures. This is where there are high local concentrations of $SiO_2$ and FeO. Internal migration of atoms will cause larger and larger parts of the particles to soften. The first internal "melts" of material will form at approximately 1100°C (1373 K) and will consist of gangue and to a large extent FeO, because at 1100°C (1373 K), the O/Fe ratio in the ferrous burden is 0.5 or lower. When gangue starts to melt, it will come into contact with the slag components of other parts of the ore burden and the slag composition will be averaged. This happens at high FeO concentrations.

The basicity of the remaining materials equals the basicity of the input materials. **Figure 8** shows a simplification of the real situation in a blast furnace and shows a slag phase diagram with the three main components: CaO, $SiO_2$, and FeO. It shows that slag melting starts at low temperatures, but that the basicity of the burden affects melting. As the FeO is reduced out of the primary melt, the slag liquidus temperature increases, but a much higher liquidus is seen for materials with a higher basicity.

Since hydrogen can easily diffuse into a more solid structure, the hydrogen reduction continues after CO reduction has stopped. When the melts are heated further and start to drip, the melt consists of a blend of the gangue, FeO, and finely dispersed iron, which has not been separated from the melt. The first process in the "primary" melt is that the gangue loses its FeO. As soon as the FeO is removed and the primary melt flows over the coke, the iron starts to dissolve carbon from the coke, which lowers the

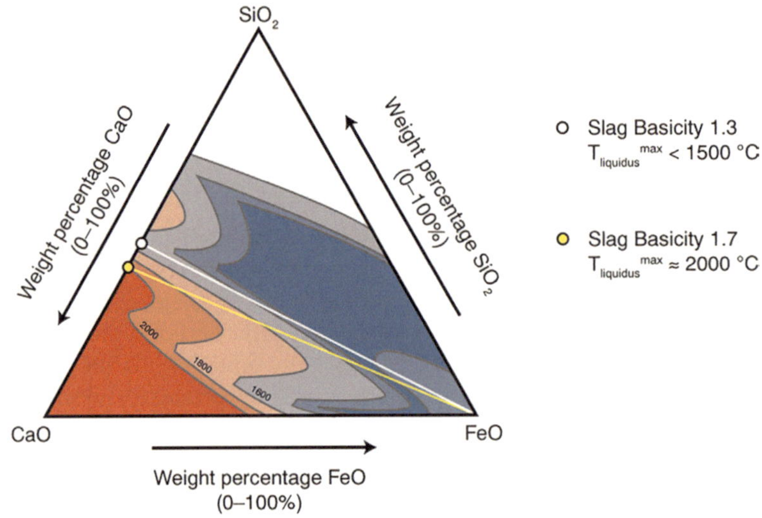


**Figure 8.**
*Slag liquidus temperatures of CaO–$SiO_2$–FeO. Slag melting starts at 60–70 wt% FeO in the structure. The yellow line indicates ferrous burden basicity (CaO/$SiO_2$) of 1.5 and the white line 1 [2].*

melting temperature rapidly. As long as the slag contains FeO, the silicon in the hot metal will be oxidized back to $SiO_2$ and the FeO in the slag reduced to Fe by coke or saturated carbon in molten iron. At the wall of the furnace, the root of the cohesive zone is located a small distance above the tuyeres, while in the center the cohesive is located higher in the furnace. As a consequence, the slag formed at the wall will have relatively high FeO and the hot metal formed at the wall will have low silicon, while the hot metal formed and dripping down in the center of the furnace will have high silicon. The content of FeO in the final slag is less than 0.5 wt%, and the total recovery rate of Fe is more than 99.7%.

## 6. Reduction of wustite by purehydrogen in future

The application of hydrogen technology provides a new idea for the steel industry to achieve the goals of carbon peaking and carbon neutralization, especially in China. Hydrogen metallurgy has become the clean, efficient, and pollution-free steelmaking technology direction of exploration in different countries.

Zhang Jianliang *et al*. have studied the reduction process of solid/liquid wustite by hydrogen using isothermal thermogravimetric analysis. The paper compared the differences between the reaction processes of hydrogen reduction of solid FeO samples and liquid FeO samples. Wustite in both states can be reduced entirely at all temperatures [3–8].

Guo Peimin et al. have promoted the development of the pure-hydrogen reduction process by the methods of theoretical analysis and calculation to discuss the industrial application prospect and key issues in basic theory and application. Heating hydrogen to the target temperature is an important factor to ensure the reduction of iron oxide. The discussion on the control step should pay attention not only to the reduction reactions but also to the heat transfer. This basic issue is directly related to the design of hydrogen reduction reactors, thus indicating the need for further studies. The important factors affecting the smooth operation of the hydrogen reduction process includes preparation of high-quality oxidized pellets, preheating of hydrogen, design of reduction device, composition and pressure of reducing gas, and security issues [9–20].

## 7. Conclusions

Based on previous research, the chapter summarizes the reduction thermodynamics and kinetics of wustite by C, CO, and $H_2$. When the temperature is lower than 810°C (1083 K), the reduction ability of CO is stronger than that of $H_2$. If it is higher than 810°C (1083 K), the opposite is true. Especially, the reduction process of wustite in blast furnace is discussed in detail. Wustite is the most difficult to be reduced in the three iron oxides. In the future, research on the reduction by pure hydrogen in ironmaking will also be the key to realize carbon neutrality. When the temperature is 1173 K, according to calculation, considering the theoretical amount and the chemical balance of the reduction reaction, the amount of hydrogen required to be put into the reactor is 952 $Nm^3$/t Fe.

## Acknowledgements

This work was funded by "Basic Research Project of Education Department of Liaoning Province, No. LJKMZ20221688", "National Natural Science Foundation of China, No. 52074086", "Entrepreneurship and Innovation Program of Enterprise and Doctor by Yingkou, No. 202116 and No. 202010", "Key R&D Project of Liaoning Province, No. 2020JH2/10100019 and No. 2022JH2/101300124". And also the special thanks to the projects, which are sponsored by Liaoning BaiQianWan Talents Program and Liaoning Innovation talents.

## Conflict of interest

The authors declare no conflict of interest.

## Appendices and nomenclature

| | |
|---|---|
| $\Delta G^{\theta}$ | standard Gibbs free energy, J/mol |
| $\Delta H^{\theta}_{298}$ | standard enthalpy change of reaction, J/mol |
| $K$ | equilibrium constant of reaction |
| $p_{CO_2}, p(CO_2)$ | partial pressure of $CO_2$ |
| $p(CO)$ | partial pressure of CO; |
| $T$ | absolute temperature, K; |
| $p_{H_2O}, p(H_2O)$ | partial pressure of $H_2O$ |
| $p_{H_2}, p(H_2O)$ | partial pressure of $H_2$ |
| $\varphi(CO_2)$ | volume fraction of $CO_2$ |
| $\varphi(CO)$ | volume fraction of CO |
| $\varphi(H_2O)$ | volume fraction of $H_2O$ |
| $\varphi(H_2)$ | volume fraction of $H_2$ |

| | |
|---|---|
| $a_{(O)}$ | activity of oxygen atom |
| $r_0$ | original radius of ore ball |
| $r_i$ | radius of unreacted core |
| $c_A^0$ | reducing gas concentration in the gas mainstream, mol/m$^3$ |
| $c_A^s$ | reducing gas concentration on the surface of the ball, mol/m$^3$ |
| $c_A^i$ | reduction gas concentration on reaction interface at any time, mol/m$^3$ |
| $c_P^i$ | product gas concentration on reaction interface at any time, mol/m$^3$ |
| $c_P^s$ | product gas concentration on the surface of the ball, mol/m$^3$ |
| $c_P^0$ | product gas concentration in the gas mainstream, mol/m$^3$ |
| $c_A^*$ | reducing gas concentration in equilibrium, mol/m$^3$ |
| $c_A$ | reducing gas concentration, mol/m$^3$ |
| $c_P$ | product gas concentration, mol/m$^3$ |
| $R_A$ | the overall reaction rate, mol/s |
| $\beta_A$ | mass transfer coefficient in boundary layer, m/s |
| $D_{eff}$ | effective diffusion coefficient of gas in micropores of the ball, m$^2$/s |
| $k_+$ | reaction rate constant, m/s |
| $f$ | reduction degree |
| $\rho_O$ | density of oxygen in the ball, kg/m$^3$ |

## Author details

Zuoliang Zhang
Liaoning Institute of Science and Technology, Benxi, China

*Address all correspondence to: zhang231167@163.com

**IntechOpen**

© 2023 The Author(s). Licensee IntechOpen. This chapter is distributed under the terms of the Creative Commons Attribution License (http://creativecommons.org/licenses/by/3.0), which permits unrestricted use, distribution, and reproduction in any medium, provided the original work is properly cited.

## References

[1] Xiaoliu W, Shengli W, editors. Ferrous Metallurgy (Ironmaking Part). 3rd ed. Beijing: Metallurgical Industry Press; 2013. 104 p; 105 p; 108-109 pp. 127-128 pp

[2] Geerdes M, Chaigneau R, Kurunov I, et al. Modern Blast Furnace Ironmaking : An Introduction. 3rd ed. Amsterdam: IOS Press; 2015. pp. 128-131

[3] Jianliang Z, Yang L, Zhengjian L, et al. Isothermal kinetic analysis on reduction of solid/liquid wustite by hydrogen. International Journal of Minerals , Metallurgy and Materials. 2022;**29**: 1830-1838. DOI: 10.1007/s12613-022-2518-0

[4] Katayama H, Taguchi S, Tsuchiya N. Reduction of iron oxide in molten slag with H 2 gas. Tetsu-to-Hagane. 1982;**68**: 2279

[5] Ban-Ya S, Iguchi Y, Nagasaka T. Rate of reduction of liquid wustite with hydrogen. Tetsu-to-Hagane. 1984;**70**: 1689

[6] Hayashi S, Iguchi Y. Hydrogen reduction of liquid iron oxide fines in gas-conveyed systems. ISIJ International. 1994;**34**:555

[7] Nagasaka T, Hino M, Ban-Ya S. Interfacial kinetics of hydrogen with liquid slag containing iron oxide. Metallurgical Materials Transaction B. 2000;**31**:945

[8] Seftejani MN, Schenk J. Kinetics of molten iron oxides reduction using hydrogen. In: International Congress on Science and Technology of Steelmaking. Venice; 2018

[9] Lei W, Peimin G, Lingbing K, et al. Industrial application prospects and key issues of the pure-hydrogen reduction process. International Journal of Minerals , Metallurgy and Materials. 2022;**29**:1922-1931. DOI: 10.1007/s12613-022-2478-4

[10] Spreitzer D, Schenk J. Reduction of iron oxides with hydrogen—A review. Steel Research International. 2019;**10**: 1900108

[11] Xu H, Zou ZS, Zhou YS, Li ZY, Yu AB. Preliminary numerical simulation of shaft furnace process for DRI production. Journal of Materials Metallurgy. 2009;**8**:7

[12] Zhao ZW, Kong FL, Tong LG, Yin SW, Xie YR, Wang L. Analysis of CO 2 emission reduction path and potential of China's steel industry under the "3060 target". Iron Steel. 2022;**57**:167

[13] Xu ZY, Wu BC. Simulating study on direct reduction of iron ore by midrex process. Gangtie. 1987;**22**:8

[14] Bai MH, Fu YX. Design method of reduction segment parameters of the gas-based direct reduction shaft furance. Journal of Iron and Steel Research. 2015; **27**:21

[15] Li F, Chu MS, Tang J, Liu ZG. Environmental impact analysis of hydrogen shaft furnace-electric furnace process. China Metallurgy. 2021;**31**:104

[16] Sarkar S, Bhattacharya R, Roy GG, Sen PK. Modeling MIDREX based process configurations for energy and emission analysis. Steel Research International. 2018;**89**:1700248

[17] Wang XN, Fu GQ, Li W, Zhu MY. Numerical simulation and optimization of flash reduction of iron ore particles

with hydrogen-rich gases. Powder Technology. 2020;**366**:587

[18] Hamadeh H, Mirgaux O, Patisson F. Detailed modeling of the direct reduction of iron ore in a shaft furnace. Materials (Basel). 2018;**11**:1865

[19] Valipour MS, Motamed Hashemi MY, Saboohi Y. Mathematical modeling of the reaction in an iron ore pellet using a mixture of hydrogen, water vapor, carbon monoxide and carbon dioxide: An isothermal study. Advanced Powder Technology. 2006;**17**:277

[20] Kazemi M, Pour MS, Du SC. Experimental and modeling study on reduction of hematite pellets by hydrogen gas. Metalluric Materials Transaction B. 2017;**48**:1114

Chapter 5

# An Innovative Process for Production of $TiO_2$ from Low Grade Raw Materials

*Dayanand Paswan and Malathi Madhurai*


## Abstract

Conventionally $TiO_2$ is produced from Ilmenite ore by carbothermic reduction followed by leaching of iron to produce high purity $TiO_2$ which leads to loss of metallic iron and disposal of effluents. Therefore, an attempt has been made to develop a process for production of $TiO_2$ utilizing low grade raw materials. Ilmenite ore pellets of an average size of 12–14 mm were prepared by mixing 2% molasses using disk pelletizer. Reduction of green ilmenite ore pellets were carried at temperature range from 1100°C to 1300°C in bed of low grade coal at a time intervals of 60 minutes from 1 to 5 hours. The activation energy for reduction of ilmenite ore pellets with lean grade coal was found to be 52.3 kJ/mole, this value corresponds to solid diffusion control reaction. Semi-empirical correlation was developed for prediction of % metallization. Reduced Ilmenite Pellets (RIP) were melted in 10 kg air as well as vacuum induction furnaces. It was observed that the total recovery of $TiO_2$ in slag was up to 85% and the separation of iron from $TiO_2$ was more than 90%.


**Keywords:** ilmenite ore, lean grade coal, pellet, $TiO_2$, induction melting, vacuum induction melting, reduction, tunnel kiln, separation of $TiO_2$

## 1. Introduction

In India around 629.57 million tons of ilmenite ore reserve including leucoxene grade is explored up to year 2016. Leucocxene is a weathered ilmenite of variable concentration but similar to pseudo-rutile Fe2Ti3O9 and called brown ilmenite. Out of 629.6 million tons of total resources; the titanium minerals are something like 394 million tons including ilmenite of 335.6 million tons, rutile of 13.4 million tons, leucoxene of 1.0 million tons, anatase of 3.3 million tons and titaniferous magnetite of 40.6 million tons. Generally, ilmenite ore is first treated chemically to obtain synthetic rutile of 90% $TiO_2$. Commercial technologies available for production of $TiO_2$ from ilmenite ore are mainly Sulphate and Chloride processes. Recently, Process Research ORTECH, Inc., Canada has developed CTL process which combines the above two processes. These processes are using high grade raw materials and an acidic media is used as a leaching agent for separation of $TiO_2$ from iron, which generates harmful gases during processing. The annual production of ilmenite, synthetic rutile and $TiO_2$ in India was around 595 thousand tons, 14.9 thousand tons and 16 thousand tons respectively in the year of 2016–2017 [1, 2]. In view of the above process deficiencies the main ambition of the present investigation is to develop an innovative process for production of $TiO_2$ utilizing lean grade raw materials via reduction technique and separation of iron from $TiO_2$ through melting route. Conventional reduction route of ilmenite ore concentrates was investigated by several researchers by mixing of carbonaceous materials such as graphite, carbon, anthracite coal, coke breeze or char, Aluminum and even with sodium compounds such as soda ash, sodium carbonate and sodium oxide to enhance the reduction efficiency and improve the metallization of iron in the ilmenite ore at diverged temperature ranges of 800 to 1200°C [3–17]. The atmosphere used for the above said processes are mainly: Ar, $N_2$, $H_2$, He, $CH_4$, $NH_3$ [10] or mixture of two gases [18, 19]. Rotary Kiln [12], TG [4, 20], EAF [2, 10], Plasma [21, 22], Microwave [23], Blast Furnace [24–26], Mechanical Alloying [27, 28], Fluidized Bed Reactor [29, 30], etc., were used as a reactor for reduction purpose. However, in the present study clay crucibles were used as a reaction chamber for reduction and electrically heated muffle furnaces are used as a source of heat. This process does not involve any inert atmosphere for reduction. The quality of coal used in the present investigation for reduction of ilmenite ore pellets has low reactivity, low fixed carbon and high ash content. Compare to the conventional processes, the present processes have following advantages which explored an opportunity to utilize the undersized ilmenite ore fines through pelletization and lean grade non coking coal for reduction:

a. No mixing of carbon/coal with ilmenite ore pellets thus restricts the diffusion of Sulfur from coal to the ilmenite ore pellets during reduction.

b. Lean grade non-coking coal containing high percentage of ash has utilized as a reductant for reduction.

c. Binder used for pelletization of ilmenite ore is less compared to conventional process to obtain optimum green and dry strength.

d. Dried ilmenite pellets were reduced in static bed of coal fines without any induration which eliminates the high temperature induration process steps.

e. As reductions were carried out in the normal atmosphere, therefore no inert gas or external reducing gas are obligatory during reduction.

Considering the above mentioned advantages, an attempt has been made in the present investigation to explore the process feasibility for reduction of ilmenite ore pellets using lean grade non coking coal and separation of $TiO_2$ from iron via melting route.

## 2. Materials and experimental

### 2.1 Raw materials used

#### *2.1.1 Ilmenite ore*

Ilmenite ore was collected from Tata Steel Ltd., Jamshedpur, Jharkhand, India and ground into −200 mesh size for finding out its chemical composition and the phases

| Elements | $TiO_2$ | FeO | $Fe_2O_3$ | Total Fe | $Al_2O_3$ | $SiO_2$ | MnO | MgO |
|---|---|---|---|---|---|---|---|---|
| % Composition | 53.18 | 15.36 | 27.51 | 31.19 | 1.06 | 1.36 | 0.3 | 0.24 |

**Table 1.**
*The chemical analysis of the ilmenite ore concentrate.*

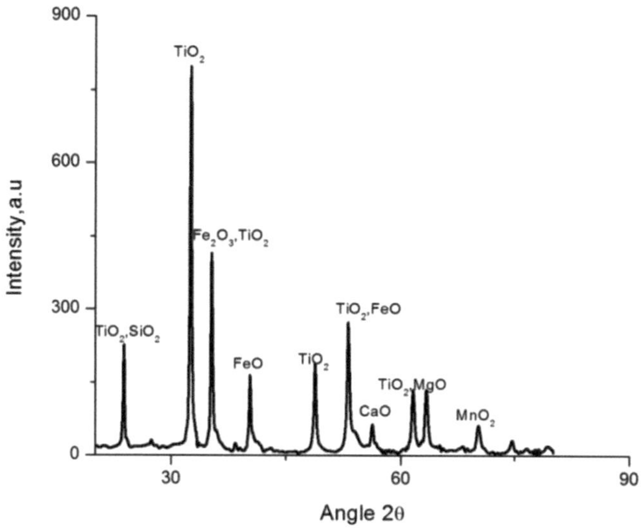


**Figure 1.**
*XRD pattern for ilmenite concentrate.*

analysis which are represented in **Table 1** and **Figure 1** respectively. The chemical analysis of ilmenite ore shows the total iron ($Fe_T$) of about 31.2 and 53.2% of $TiO_2$ in the form of $Fe_2Ti_3O_9$ and $FeTiO_3$ phases which was confirmed by XRD analysis.

### *2.1.2 Lean grade non-coking coal fines*

Lean grade non-coking coal fines used for present study were collected from Tata Steel Ltd., Jamshedpur, Jharkhand, India and ground into −200 mesh size for its chemical characterization. The composition of the lean grade non-coking coal fines is charted in **Table 2**. The fixed carbon present in the lean grade non-coking coal fines was 58.2% C having the property of low reactivity and the volatile matter in the coal was around 9.46%. The ash content present in the lean grade coal fine was about 31%.

### *2.1.3 Binder*

Commercially available grade of molasses was used in the present investigation for binding of ilmenite ores to make pellets.

| Coal | %C | %VM | %Moisture | %S | %Ash |
|---|---|---|---|---|---|
| Low grade | 57.72 | 9.46 | 0.91 | 0.28 | 31.91 |

**Table 2.**
*Proximate analysis of low grade coal.*

## 2.2 Experimental procedures

### *2.2.1 Preparation of ilmenite ore pellets*

Ilmenite ore pellets of 12–16 mm diameter were made in a laboratory scale disc pelletizer utilizing −72 mesh size ilmenite ore fines with 2 wt. % molasses as a binding agent in presence of moisture. In order to remove excess moisture, the drying of green ilmenite ore pellets was performed in the normal atmosphere for 24 hours. The diameter and weight of the dried ilmenite ore pellets were recorded for further studies. **Figure 2** shows the green ilmenite ore pellets made in a laboratory scale disc pelletizer.

The air dried green ilmenite ore pellets were subjected to a compressive strength machine to obtain the green strength for further handling and processing. The handling strength required for reduction of ilmenite pellets in the static bed is about 6–8 drops which is much less than the conventional drop strength of pellets required for rotary kiln process.

### *2.2.2 Reduction procedures*

Reduction experiments of air dried green ilmenite ore pellets were carried out with lean grade coal fines in an electrically heated muffle furnace in a batch of 100 grams by simulating the conditions of a tunnel kiln [31]. Lean grade coal fines was ground, screened to −50 mesh size and charged in a 60 mm height 25 mm inside diameter clay crucibles in a predesigned fashion along with the green ilmenite ore pellets. The amount of lean grade coal used for present study was 20% over the stoichiometric requirement. Pt - 10% pt-Rh thermocouple was used to measure the temperature of the furnace. A Schematic view of the experimental apparatus used for the present investigation is shown in **Figure 3**. The percentage reduction of ilmenite ore pellets was calculated as the amount of oxygen removed during reduction and estimated by mass balance as given in Eq. (1).

$$\text{Degree of Reduction (\%R)} = \frac{\text{Loss in wt. of pellet} - \text{LOI}}{\text{Wt. of oxygen in the pellet}} \times 100 \qquad (1)$$

where LOI is the Loss of Ignition,

**Figure 2.**
*Green ilmenite ore pellets.*

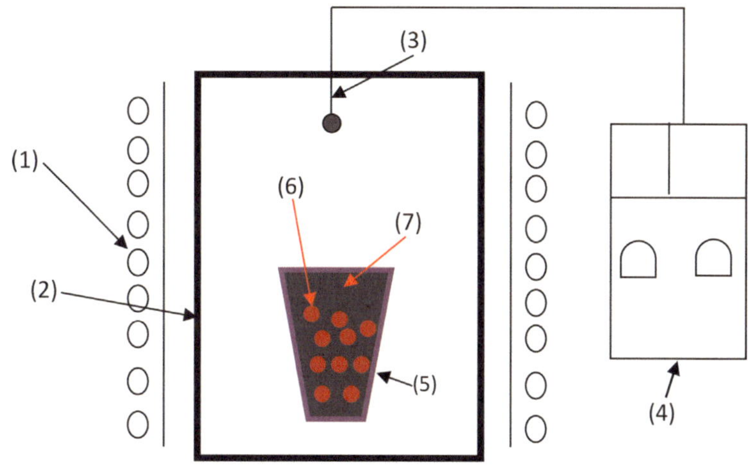


**Figure 3.**
*Schematic diagram of the experimental apparatus includes: (1) heating element, (2) muffle furnace, (3) thermocouple, (4) temperature controller, (5) clay crucible, (6) ilmenite pellets and (7) lean grade coal [31].*

The factors which are influencing the degree of reduction of ilmenite pellets are: (a) chemical composition of raw materials (ilmenite ore and lean grade coal), (b) particle size of ilmenite ore, (c) particle size of low grade coal (d) shape and size of ilmenite ore pellets (e) fraction/amount of binder, (f) percentage moisture with constant amount of binder (2 wt% of molasses) on green strength of the ilmenite ore pellets (g) amount of coal, (h) temperature and (i) time. Ilmenite ore used for this investigation contains around 40–50% $TiO_2$. Therefore, carbothermic reduction is feasible for reduction of such a high iron content in ilmenite ore by suitable thermodynamic reactions and conditions to produce high grade pig iron as a valuable byproduct. During course of reduction, the ilmenite ore pellets undergoes through various types of direct and indirect reduction before converting into the metallic iron and $TiO_2$. The various types of chemical reactions during reduction process are given in **Table 3**.

| Sl. No. | Reactions | Remarks |
|---|---|---|
| 1. | $C + O_2 = CO_2$ | Gasification of carbon |
| 2. | $C + CO_2 = 2CO$ | |
| 3. | $C + ½ O_2 = CO$ | |
| 4. | $C + H_2O = CO + H_2$ | |
| 5. | $3Fe_2O_3.TiO_2 + CO = 2Fe_3O_4.TiO_2 + CO_2$ | Indirect reduction |
| 6. | $Fe_3O_4. TiO_2 + CO = 3FeO.TiO_2 + CO_2$ | |
| 7. | $FeO.TiO_2 + CO = Fe.TiO_2 + CO_2$ | |
| 8. | $Fe_2O_3.TiO_2 + 3CO = 2 Fe.TiO_2 + 3CO_2$ | |
| 9. | $2Fe_2O_3.TiO_2 + 3C = 4 Fe.TiO_2 + 3CO$ | Direct reduction |

**Table 3.**
*Reactions taking place during reduction of ilmenite ore pellets.*

However, for an industrial scale production, the shape and optimum size of the pellets are spherical and are about 12–16 mm diameter respectively is needed for an effective reduction or DRI making process. Furthermore, the particle sizes used for ilmenite ore and lean grade coal are −72 mesh and − 50 mesh respectively. Therefore, the author has not considered the effect of ilmenite ore pellet size on degree of reduction. The above mentioned properties of materials are important on reduction behavior of ilmenite ore pellets; but the main driving force required for reduction is amount of coal, temperature and time. Therefore, in the following section of results and discussion the author would try to explain first the effect of percentage moisture with constant amount of binder (2 wt% of molasses) on green strength of the ilmenite ore pellets. Secondly, the reduction behavior of ilmenite ore pellets was studied in terms of temperature, time, and percent metallization. Thirdly, the author would try to compare reduction behavior of ilmenite ore pellet with normal iron ore pellets and thereof an empirical relation. At last, the effect of change of amount of low grade coal on bulk (500 gram of scale) reduction of ilmenite ore pellets were performed to ensure the reproducibility of results at mass scale.

## 3. Results and discussion

### 3.1 Effect of moisture on green strength of ilmenite ore pellets

The effect of percentage moisture with constant amount of binder (2%wt. of molasses) on green strength of the ilmenite ore pellets were measured by known method [32]. It can be noted that the green compressive strength of the ilmenite pellets are first increases with addition of moisture content and then start decreasing. With the added water, where maximum strength obtained was considered to be the optimum moisture for pelletization, which is depicts in **Figure 4**. It was observed that maximum green strength of the ilmenite pellets was obtained at 12 wt. % of water

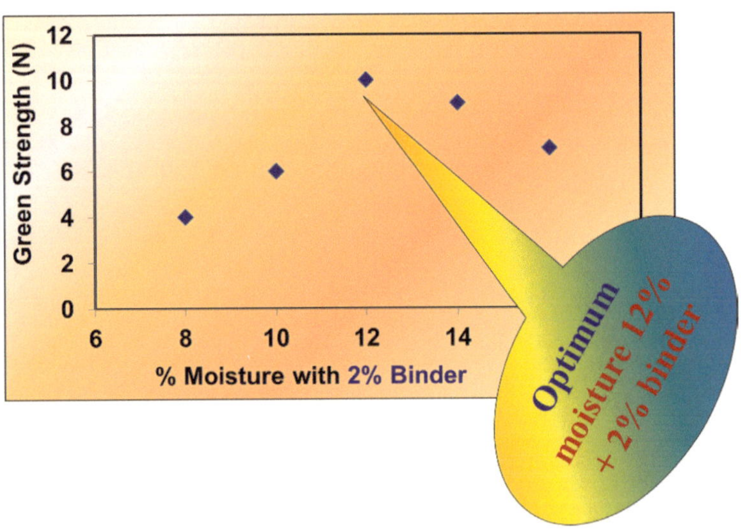


**Figure 4.**
*Effect of moisture on green strength of ilmenite ore pellets.*

(with 2 wt. % of molasses) due to complete void filling which provide maximum capillary action.

### 3.2 Effect of temperature on reduction of green ilmenite ore pellets

Series of reduction experiments were performed in the temperature range of 1100–1300°C. Reduction time of the samples were varied from 60 minutes to 360 minutes. In order to facilitate the collection of samples at a time interval of 60 minutes, six crucibles containing samples were simultaneously inserted in the furnace and reduced at constant temperature by simulating Tunnel Kiln condition at laboratory. This assisted in monitoring the progress of reduction as well as to examine the physical condition of the reduced pellets as a function of time. Precautions were exercised to minimize re-oxidation of the reduced pellets after taking out from the reaction chamber at hot condition and while cooling. To ensure complete reduction of ilmenite pellets, the amount of carbon in the lean grade coal used was 20% over the stoichiometric calculation. **Figure 5** represents the percent reduction against time at the six experimental temperatures. Percent reduction was calculated in the usual manner as the removed oxygen to the removable oxygen. Multiple pellets were used to calculate the average percent reduction at all the time and temperatures. It is evident from **Figure 5** that the percent reduction of ilmenite pellets keeps on increasing with increase of time and temperature. It is also observed that the percent reduction of ilmenite pellets at 1300°C is completed between 3 to 4 hours of reduction time. The increase in the reduction time above this temperature leads to further growth and fusion of their own whisker within their reduced ilmenite pellets.

### 3.3 Effect of %metallization of reduced ilmenite pellets

After reduction experiments the reduced ilmenite pellets were separated from the coal bed and analyzed for their composition and % metallization. Percent

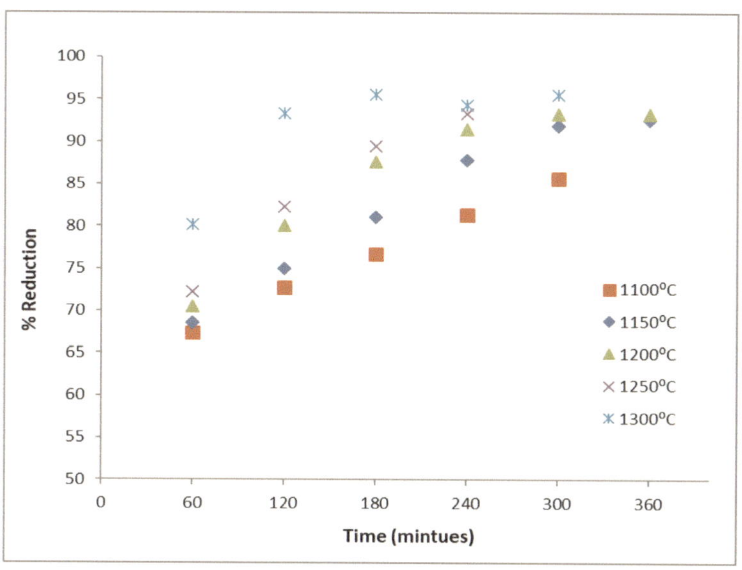


**Figure 5.**
*Reduction behaviors of ilmenite ore pellets in bed of lean grade coal.*

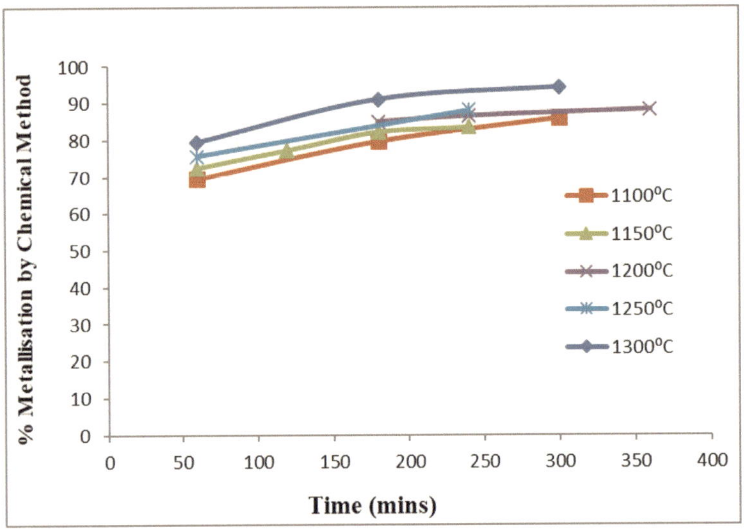


**Figure 6.**
*Percent metallisation of reduced ilmenite ore pellets in bed of low grade coal.*

metallization as a function of reduction time and temperature is shown in **Figure 6**. As the time and temperature increases the % metallization also increases because of fusion of iron particles in the matrix of ilmenite.

## 3.4 Correlation between % reduction and % metallisation during reduction of ilmenite ore pellets in the bed of low grade coal

The correlation between percent reduction and percent metallisation of ilmenite ore pellets reduced in the bed of low grade coal has been depicted in **Figure 7** by

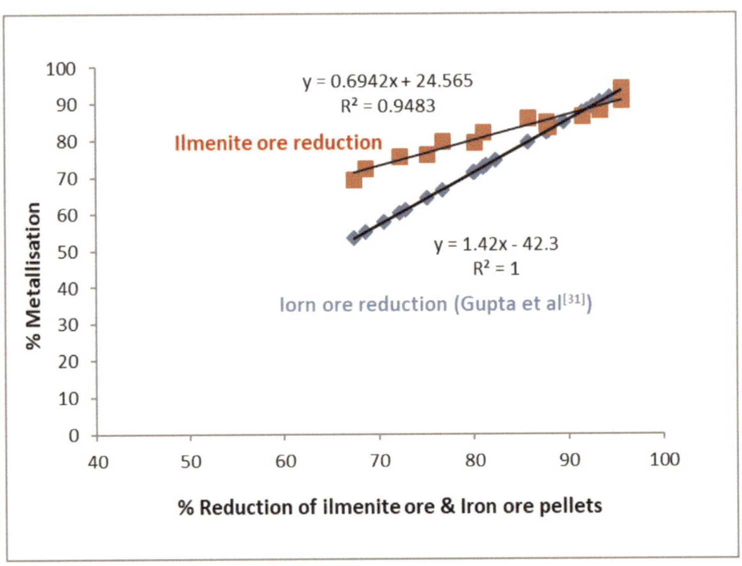


**Figure 7.**
*Relation between % reduction and % metallisation during reduction of ilmenite ore pellets in bed of low grade coal.*

simple regression fit analysis. It is clear from **Figure 7** that ilmenite ore reduction in the initial period is much faster than iron ore reduction and also it is clear that achieving high metallization percent is very difficult as the time proceeds. It is very similar to the empirical relation developed by Gupta et al. [32] for the case of iron ore pellet reduction. In the case of ilmenite ore pellet reduction, the percent metallization can be predicted using the following empirical Eq. (2).

$$\%\text{Metallisation} = 0.694\%\text{Reduction} + 24.56 \tag{2}$$

### 3.5 Reduction kinetics of ilmenite ore pellets embedded in the low grade coal

For diffusion controlled kinetics the following Eq. (3) is applied for solid reaction on the reduction of ilmenite ore pellets on the bed of low grade coal.

$$\left(1 - [1 - f]^{1/3}\right)^2 = kt \tag{3}$$

Where, f, is the fractional reduction, t is the time (minutes) and k is the rate constant which is slope of time (t) Vs $\left(1 - [1 - f]^{1/3}\right)^2$ plot shown in **Figure 8(a)**. **Figure 8(b)** shows the relation between lnK and 1/T for reduction of ilmenite ore pellets in the bed of low grade coal. The slope of logarithmic rate constant versus 1/T plot gives the activation energies of the reaction which is calculated to be 52.3 kJ/mole. This value is corresponding to solid diffusion control reaction while reduction of ilmenite ore pellets in the bed of low grade coal.

### 3.6 Physical appearance of the reduce ilmenite pellets (RIP's)

Physical appearance of cross sectioned RIP's at different temperature and time is shown in **Figure 9**. It can be observed that as the reduction temperature increases the surfaces of the pellets become more and more bright.

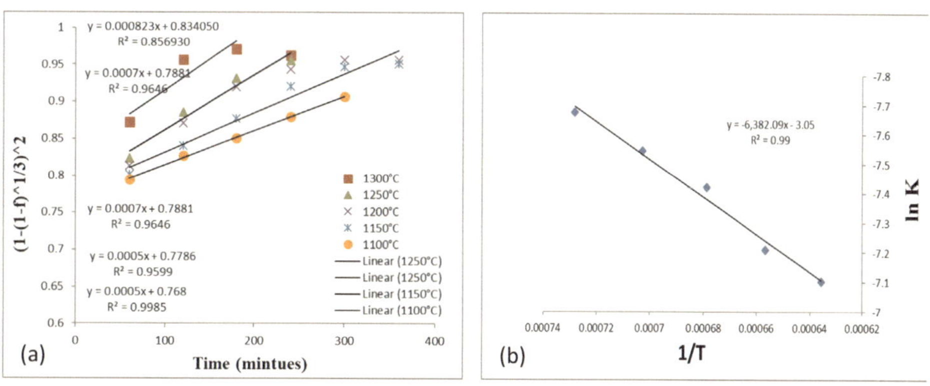


**Figure 8.**
*Reduction kinetics of ilmenite pellets reduced in the bed of low grade coal (a) relationship between diffusion controlled kinetics and reduction time (b) relation between lnK and 1/T.*

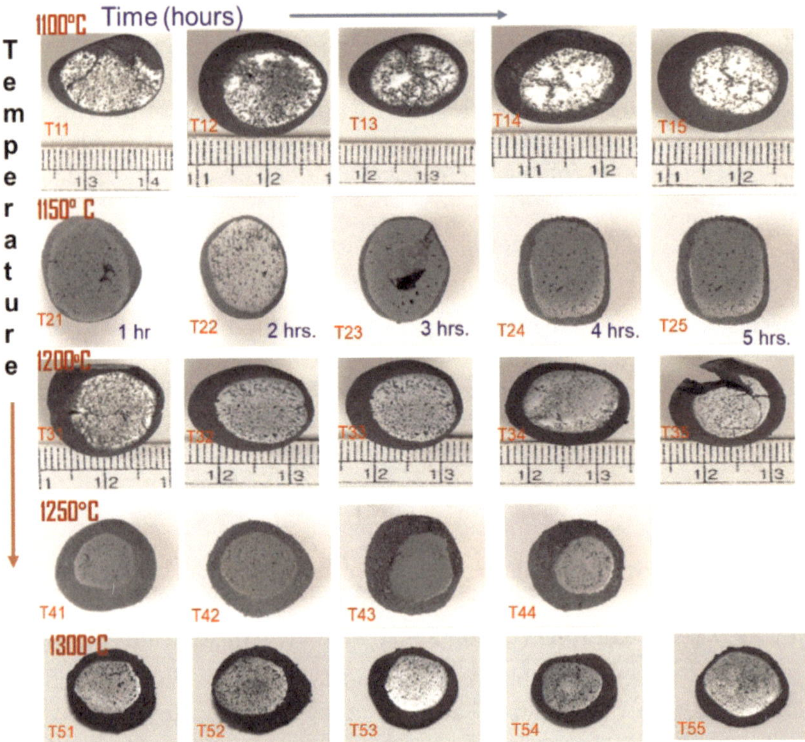


**Figure 9.**
*Physical appearance of cross sectioned reduced ilmenite pellets at different time and temperature.*

This is due to progress of conversion of iron oxide in to metallic iron with progress of reduction time and temperature. With increase in reduction time at particular reduction temperature iron whiskers grow and fused leading to metallic luster in the briquettes. The same trend is observed as the reduction temperature increase for the particular holding time.

### 3.7 Optical microstructure of reduced ilmenite pellets (RIP's)

Optical microscopy provides information on the various phases present in terms of size, dissemination, intergrowths and relative proportions. The microscope used was Leica DM 2500 M at CSIR-NML, Jamshedpur. **Figure 10** shows the optical micrographs of ilmenite pellets reduced for 210 minutes at (a) 1100°C, (b) 1150°C, (c) 1200°C, (d) 1250°C and (e) 1300°C at 5X magnification to feel the micro structure of RIP's. The reduced pellets on viewing under the light microscope showed that they are mainly composed of metallic iron (white portions) in a $TiO_2$ and gangue (black portions) matrix. It is quite evident from the micrographs that, as temperature increase the amount of metallic iron increases showing better reduction.

### 3.8 SEM-EDAX of RIP's

To confirm the presence of $TiO_2$ and gangue minerals as a matrix around the metallic iron SEM-EDAX analysis of the reduced sample were performed. The

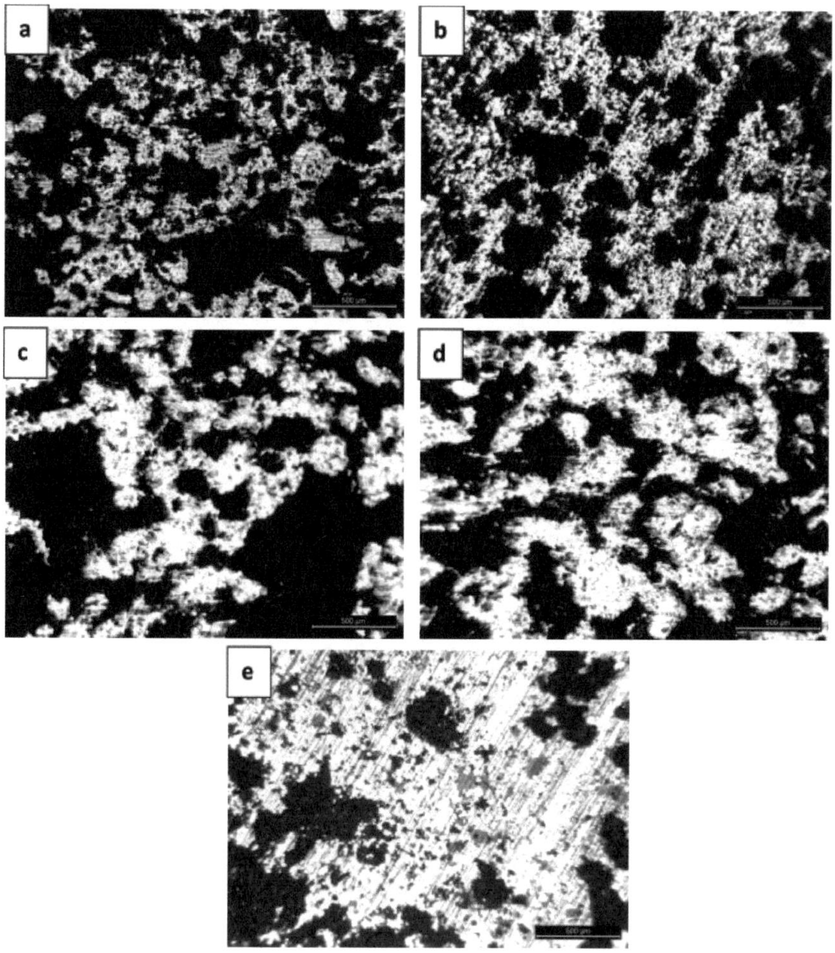

**Figure 10.**
*Optical micrographs of RIP's for time of 210 minutes at different temperatures of (a) 1100°C, (b) 1150°C, (c) 1200°C, (d) 1250°C and (e) 1300°C at 5X magnification.*

advantage of scanning electron microscopy over other techniques is that it allows a complete view of the fractured surface of the pellets due to its high depth of focus. It was observed that as temperature increases, nucleation of metallic iron takes place over the ilmenite ore particles. Number of metallic iron nuclei increased with increase in reduction temperature.

**Figure 11** shows the SEM images of ilmenite pellets reduced at different temperatures varying from 1100–1300°C for 240 minutes (a) at 100X, (b) at 500X and (c) at 1000X. It is quite evident from **Figure 11** is that as the temperature increases the formation of iron droplets are found to be more and also growth of iron whisker is predominant.

At 2000X magnification SEM-EDAX analysis of ilmenite ore pellets reduced at 1200°C for 4 hours is shown in **Figure 12**. The spot analysis of SEM at spot 2 and 4 are seems to be bulk phase which is mixture of Fe and $TiO_2$ phase but spot 1 and 3 are shows spherical shapes particle which is rich in Fe. It is quite evident from EDAX images are given in side by in **Figure 12** which is shown below.

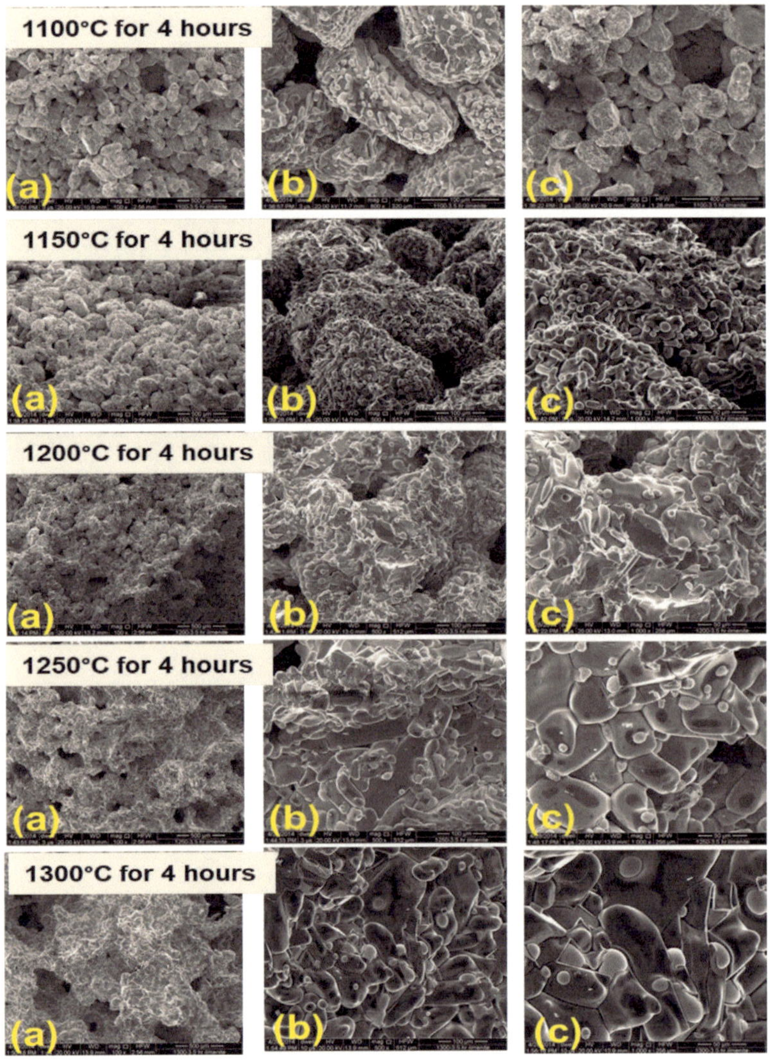


**Figure 11.**
*SEM images of ilmenite pellets reduced at 1100–1300°C with 50°C interval for 240 minutes' time (a) at 100X, (b) at 500X and (c) at 1000X.*

## 3.9 Bulk (500 gram of scale) reduction of ilmenite ore pellets and effect of amount of low grade coal on % reduction

Low grade coal used in this investigation was able to achieve % reduction of around 90–95% at high temperatures of 1200–1300°C when compared to other reducing agents like coking coal. Therefore, an attempt was made to increase the percentage reduction with various proportions of low grade coal with petroleum coke. Bulk sample of 500 grams, ilmenite ore pellets per batch were made and reduced at 1300°C for 180 minutes by varying amount of reducing agent by 20% interval. The results were given in **Table 4** and plotted in **Figure 13**. It is clearly observed from the **Figure 13** that % reduction can be improved from 95 to 98.7% by replacing low grade

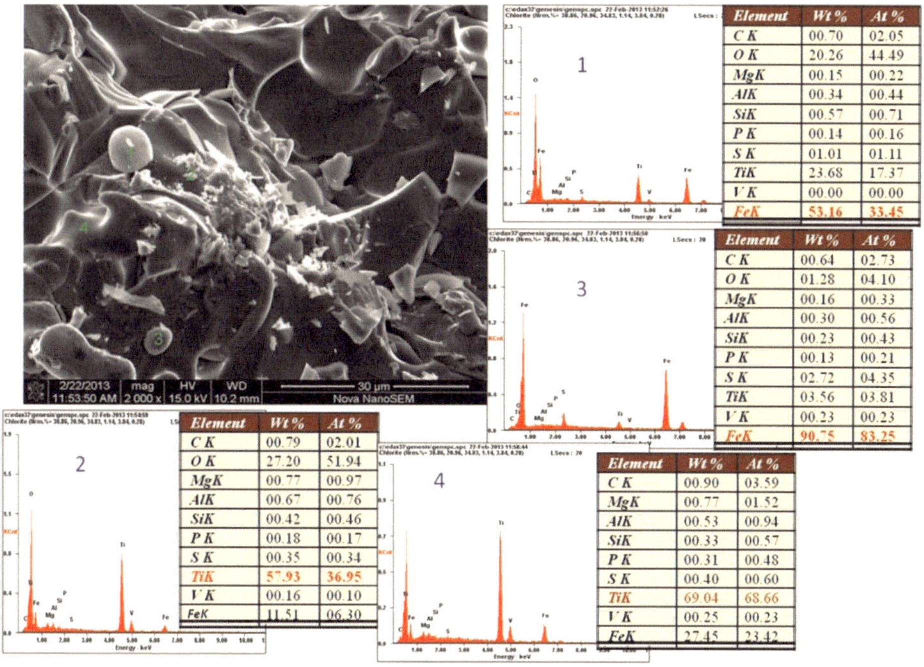


**Figure 12.**
*SEM-EDAX of ilmenite pellets reduced at 1200°C for 4 hours at 2000X magnification.*

| Sl. No. | Initial Wt. of pellets (g) | Wt. of low reactive coal (g) | Wt. of Pet. Coke (g) | Final Wt. of pellets (g) | Loss in wt. (g) | % Reduction | %Met. From Eq. (1) |
|---|---|---|---|---|---|---|---|
| 1 | 500 | 600 | 0 | 444.5 | 55.5 | 95.17 | 87.64 |
| 2 | 500 | 500 | 0 | 446.3 | 53.7 | 92.08 | 84.39 |
| 3 | 500 | 400 | 100 | 447.5 | 52.5 | 90.02 | 82.23 |
| 4 | 500 | 300 | 200 | 448.5 | 51.5 | 88.31 | 80.42 |
| 5 | 500 | 250 | 250 | 448 | 52 | 89.16 | 81.33 |
| 6 | 500 | 100 | 400 | 443.8 | 56.2 | 96.37 | 88.9 |
| 7 | 500 | 0 | 500 | 442.4 | 57.6 | 98.77 | 91.43 |

**Table 4.**
*Variation of % reduction with different combinations of reducing agents.*

coal with pet coke. But 20% extra addition of low grade coal over the stoichiometric condition would be sufficient enough to get 95% reduction and it is quite evident in **Figure 13** and also the reduction process is not demanding any metallurgical coke.

## 4. Melting and separation of $TiO_2$ from reduced ilmenite pellets

Melting and separation of Iron from reduced ilmenite pellets were carried out in 2–3 kg scale for the first time in an induction furnace. The bulk reduction of

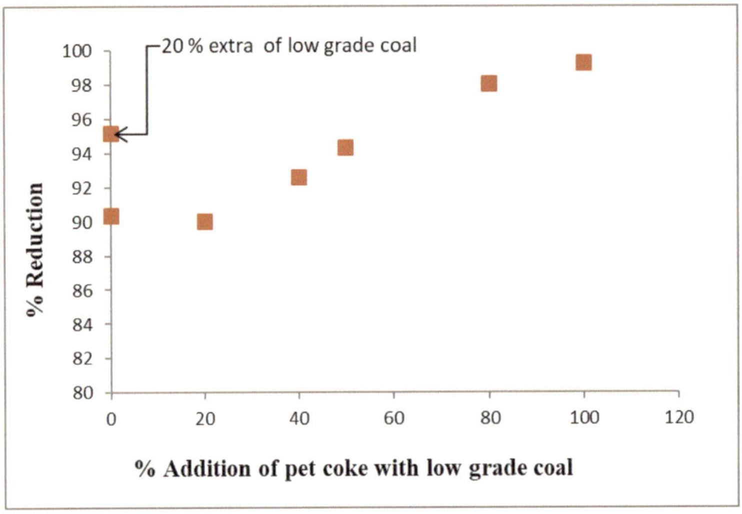


**Figure 13.**
*Effect of reducing agent with different % of coal mixed with petrollium coke.*

ilmenite ore pellets was carried out at laboratory scale muffle furnace using industrial size crucible at 2 to 3 kg by simulating tunnel kiln conditions. Here, the author deliberately used tunnel kiln conditions for reduction to scale-up of process in future has been considered for commercial production. The melting experiments were carried out in an induction furnace under normal atmosphere as well in vacuum conditions to visualize the optimum recovery of $TiO_2$. Under five different conditions (Case 1 to Case 5) melting and separation process was carried out.

In Case −1, melting was with graphite crucible of capacity 10 kg (50 kW) in an induction melting furnace. In order to dissolve/melt the reduced ilmenite pellets (RIP) in an induction furnace, 2 kg cast iron was liquefied first to make the metallic pool. Then, 1.34 kg of reduced ilmenite pellet was added gradually in the batch of 500 grams at a time. After addition of 1.34 kg RIP few minutes was given for dissolution to occur and then slag was removed before pouring the melt in to cast iron mold. The quantity of solidified slag was around 800 gms. It was separated manually as three different types: (i) Fully magnetic (140 g) (ii) semi-magnetic (640 g) and (iii) non-magnetic (20 g) and send for standard chemical analysis. A typical melting procedure is given in **Figure 14**. The **Table 5** is representing the chemical analysis of three different slag samples obtained from the experiment. From the **Table 5,** it is observed that the non-magnetic slag samples are having around 62% $TiO_2$ whereas semi-magnetic slag samples contain only 59% $TiO_2$.

### 4.1 Case: 2–Air induction melting with fluxing agent

The crucible capacity used for experiment is about 5 kg (50 kW) in an induction melting furnace. In this case reduced ilmenite pellets of 100 grams, $Na_2CO_3$ flux of 60 grams and 1.0 kg of cast iron was used as charge material. It was observed that the slag samples are magnetic in nature and containing around 57% $TiO_2$. The separation of $TiO_2$ from the melt was seems underprivileged.

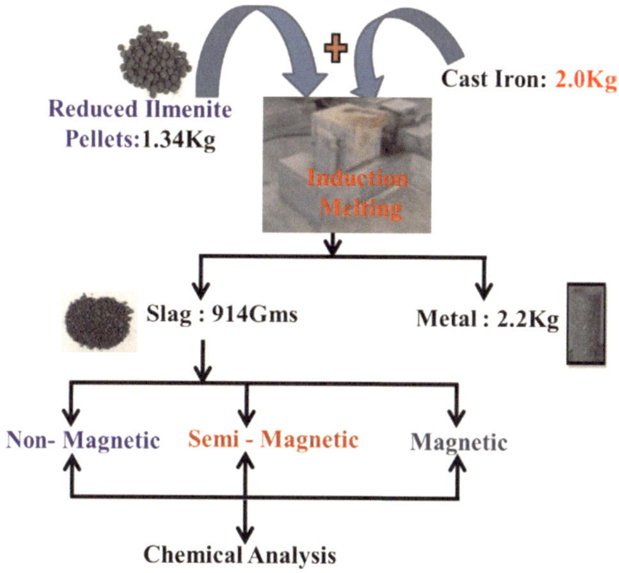


**Figure 14.**
*Case – 1: air induction melting experiment with reduced ilmenite pellets.*

| % Radicals | Chemical analysis of ilmenite slag | | |
|---|---|---|---|
| | Magnetic | Semi Magnetic | Non-Magnetic |
| $TiO_2$ | 29.91 | 59.33 | 61.41 |
| $Fe_M$ | 3.37 | 0.55 | 0.51 |
| $Fe_T$ | 47.37 | 14.6 | 16.6 |
| FeO | 33.87 | 12.2 | 8.31 |
| $Fe_2O_3$ | 25.28 | 6.52 | 13.84 |
| MgO | 1.10 | 0.83 | 0.7 |
| $SiO_2$ | 2.19 | 1.56 | 1.98 |
| $Al_2O_3$ | 0.97 | 1.33 | 1.44 |
| C | 0.54 | NF | NF |

**Table 5.**
*Chemical analysis of slag sample collected during air induction melting.*

## 4.2 Case: 3–Vacuum induction melting

For the first time, the melting of Reduced Ilmenite Pellets (RIP) was carried out with graphite crucible of capacity 10 kg in a vacuum induction furnace. Magnesia refractory materials were used as a lining to avoid the unseen attack of the slag on the furnace wall. The feed consists of 3.5 kg cast iron and 800 gms of RIP was used in this experiment. It was observed that the slag samples containing 58% $TiO_2$ along with 30% MgO, 8% $SiO_2$ and total Fe 1.1%. Furthermore, it has been observed from the chemical analysis that about 40% is MgO + $SiO_2$, this may be due to erosion of magnesia lining during melting. If we avoid the dissolution of the lining during melting, the total recovery of $TiO_2$ in slag may reach up to 85%. The total slag collected = 700 g (i.e. lining material + slag) but everything was non-magnetic in nature. In this case, $TiO_2$ slag was having more than 85% purity.

### 4.3 Case: 4–Vacuum induction melting

The feed consists of 2 kg cast iron and 1.5 kg of RIP's. It was observed that the peephole glass of the furnace started getting fogginess. This fogginess was due to evaporation of the moisture inside the furnace and may be due to vaporization of some of elements of the charge material.

### 4.4 Case: 5–Reduction of ilmenite pellets in air induction furnace

An attempt has been made to reduce the ilmenite pellets in the bed of lean grade coal in an air induction furnace at 1400°C for 210 minutes. It was observed that the iron oxides getting reduced to metallic iron, molten and moved out of the pellets to form a pure Fe metal droplet of 90 gram leaving behind TiO2 and other gangue particles in the shell. Thus, TiO2 along with other gangue minerals got separated from iron metal.

## 5. The "iron" mass balance in the melting process

To assess the mass balance of iron in the process, the material collected for 5 different conditions (Case 1–5) were weighed and presented in **Table 6**. The iron content in the material collected from the melting experiments was estimated through chemical analysis.

From **Table 5** it is observed that the total recovery of $TiO_2$ in slag was up to 85%. The total slag collected was about 700 g (i.e. lining material + slag) but everything was in non-magnetic in nature. $TiO_2$ slag was having more than 85% purity. The yield of iron was observed to be around 83%. The feasibility of separation of iron from ilmenite ore matrix was considerably good with the use of low grade coal at high temperature. Thus, by using this technique, separation is possible at 1400°C compared to conventionally used temperatures (1650–1700°C). It is a better process to

| Melting and separation technique | Input charge material (kg) | | | Output (kg) | | | % Recovery of Iron (purity of $TiO_2$) |
|---|---|---|---|---|---|---|---|
| | Cast Iron | RIP | Flux | Slag | Metal | Fe in RIP | |
| Case - 1 | 2 | 1.34 | — | 0.81 | 2.2 | 0.402 | 50.0 (61.4%$TiO_2$) |
| Case - 2 | 1 | 0.1 | 0.06 | 0.06 | 0.85 | 0.030 | -NF (57%$TiO_2$) |
| Case - 3 | 3.5 | 0.80 | — | 0.7* | 3.7 | 0.240 | 83 (85% $TiO_2$) |
| Case - 4 | 2 | 1.5 | — | | | 0.450 | NF |
| Case - 5 | — | 0.40 | — | 0.31 | 0.089 Pure Fe | 0.121 | 73 (90% $TiO_2$) |

*Slag was mixed with refractory material.*

**Table 6.**
*Mass balance of iron in the process.*

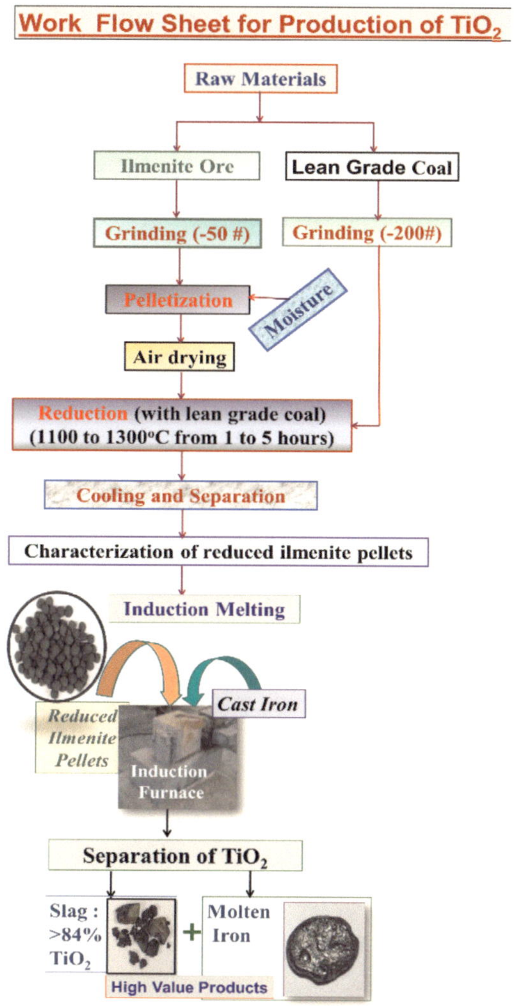


**Figure 15.**
*The overall work flow sheet for production of $TiO_2$.*

save a lot of energy. The separation of iron from ilmenite ore was quite appreciable. The overall work flow sheet for production of TiO2 is given in **Figure 15**.

## 6. Executive summary

Preliminary investigations were carried in a laboratory scale muffle furnace, simulating a tunnel kiln condition, to produce reduced ilmenite pellets from air dried Ilmenite ore pellets and lean grade coal. Reduced ilmenite pellets were found to undergo 95% of reduction at 1300°C temperature and at 3 hours of time. Activation energy required for the reaction was found to be 52.3 kJ/mole. A semi-empirical correlation was developed to predict % metallization of ilmenite ore pellets and lean grade coal of particular grade of raw materials. It was observed that reduction of ilmenite ore pellet was much faster in the initial period, as the time progress it was observed to be very difficult for achieving high percentage metallization. It was quite evident from the optical and SEM-EDAX images that as the temperature and time increases the amount of metallic iron islands increases. On heating at 1400°C for 3 hours, iron nuggets got separated as a metal droplet leaving behind $TiO_2$ and gangue as slag. Purity of slag so produced was observed to be 90% $TiO_2$. Through vacuum induction melting results shows that the total recovery of $TiO_2$ in slag was up to 85% and the yield of iron was observed to be around 83%. Overall, reduction and separation of iron from ilmenite ore pellets through tunnel kiln and induction melting appears to provide an attractive way for utilization of lean grade raw materials.

## Author details

Dayanand Paswan and Malathi Madhurai*
CSIR-National Metallurgical Laboratory, Jamshedpur, India

*Address all correspondence to: malathi@nmlindia.org

IntechOpen

© 2023 The Author(s). Licensee IntechOpen. This chapter is distributed under the terms of the Creative Commons Attribution License (http://creativecommons.org/licenses/by/3.0), which permits unrestricted use, distribution, and reproduction in any medium, provided the original work is properly cited.

## References

[1] Government of India, Ministry of Mines. Indian minerals yearbook 2019 (part III). In: Ilmenite and Rutile Advance Release. 58th ed. Nagpur: Indian Bureau of Mines, Indira Bhavan, Civil Lines; Jan 2015

[2] Murty CVGK, Upadhyay R, Asokan S. Electro smelting of ilmenite for production of $TiO_2$ slag-potential of India as a global player. In: INFACON XI, Conference Held on 18–21, Feb 2007. New Delhi, India: The Indian Ferro Alloy Producers Association, Tata Steel; 2007

[3] Yu-ming W, Zhang-fu Y, Zhan-cheng G, Qiang-qiang T, Zhao-yi L, Wei-zhong J. Reduction mechanism of natural ilmenite with graphite. Transactions of the Nonferrous Metals Society of China. 2008;**18**: 962-968

[4] Kucukkaragoz CS, Eric RH. Solid state reduction of a natural ilmenite. Minerals Engineering. 2006;**19**:334-337

[5] Suresh KG, Rajakumar V, Grieveson P. Kinetics of reduction of ilmenite with graphite at 1000 to 1100°C. Metallurgical Transactions B. 1987;**18**(B):713-717

[6] Wouterlood HJ. The reduction of ilmenite with carbon. Journal of Chemical Technology and Biotechnology. 1979;**29**:603-618

[7] EL-Guindy MI, Davenport WG. Kinetics and mechanism of ilmenite reduction with graphite. Metallurgical and Materials Transactions. 1970;**1**: 1729-1734

[8] Suresh KG, Rajakumar V, Grieveson P. The influence of weathering on the reduction of ilmenite with carbon. Metallurgical Transactions B. 1989;**18**(B):735-745

[9] Nuilek K, Memongkol N, Niyomwas S. Production of titanium carbide from ilmenite. Songklanakarin Journal of Science and Technology. 2008;**30**(2): 239-242

[10] Gueguin M, Cardarelli F. Chemistry and mineralogy of Titania-rich slags. Part 1- Hemo-ilmenite, sulphate and upgraded titania slags. Mineral Processing & Extractive Metallurgical Reviews. 2007;**28**:1-58

[11] Gawad HHA, El-Hussiny NA, Wassef MA, Khalifa MG, Soliman Aly AA, Shalabi MEH. Reducibility study of Egyptian ilmenite ore briquettes and powder with coke breeze at 800-1100°C. Indian Institute of Chemical Engineers. 2012;**54**(2):125-136

[12] Sai PST. Evaluation of mathematical models for the reduction of ilmenite with char in a rotary reactor. Industrial and Engineering Chemistry Research. 2008; **50**(4):312-322

[13] Burja J, Tehovnik F, Lamut J, Knap M. Alumothermic reduction of ilmenite in a Steel melt. Materials Technology. 2013;**47**(2):217-222

[14] Olay M, Dalane Hauge I. Process for treating ilmenite containing materials to produce metallic iron concentrates and titanium dioxide concentrates. 29 Oct 1957. US 2811434A

[15] Nafeaa IA, Zekry AF, Khalifa MG, Farag AFB, El-Hussiny NA, Kel H, et al. Sodium titanates formation by roasting of pellets or powder mixture of soda ash and rossetta region ilmenite ore

concentrate in air. Open Journal of Metal. 2014;**4**:20-30

[16] El-Tawil SZ, Morsi IM, Yehia A, Francis AA. Alkali reductive roasting of ilmenite ore. Canadian Metallurgical Quarterly. 1996;**35**:31-37

[17] Morsi IM, Abdullah FHA, Mohamed OA, Shalabi MEH, El-Tawil SZ. Processing of ilmenite ore by sintering/roasting technique. In: 3rd Mining, Petroleum and Metallurgy Conference, Cairo, 2–4 February 1992. Neue Hutte (Germany). 1992

[18] Wang Y, Yuan Z, Matsuura H, Tukihashi F. Reduction extraction kinetics of titania and iron from an ilmenite by H2-Ar gas mixtures. ISIJ International. 2009;**49**(2):164-170

[19] Rezan SA, Zhang G, Ostrovski O. Carbothermal reduction and nitridation of ilmenite concentrates. ISIJ International. 2012;**52**(3):363-368

[20] El-Hussiny NA, Lasheen TA, Shalabi MEH. Kinetic reduction of rosetta ilmenite with coke breeze and beneficiation of the product. The Journal of Ore Dressing. 2008;**10**(20):16-23

[21] Galgali RK, Ray HS, Chakrabarti AK. A Study on carbothermic reduction of ilmenite ore in a plasma reactor. Metallurgical and Materials Transactions B. 1998;**29**(B):1175-1180

[22] Samokhin AV, Alexeev NV, Sinaiskiy MA, Fadeev AA, Tsvetkov YV, Kolesnikov AV. Synthesis of Fe-Ti(O,C,N) nanosized composite via reduction of ilmenite by methane in nitrogen thermal plasma flow. In: Online-Proceedings of ISPC Conferences. Moscow, Russia: Institute of Metallurgy and Material Science; 2013. Available from: www.ispc-conference.org/ispcproc/ispc21/ID109.pdf

[23] Peng J, Yang, Huang M, Huang M. Microwave-assisted Reduction and Leaching Process of Ilmenite. Changsha, China: IEEE; 2008. pp. 1383-1386. DOI: 10.1109/ISAPE.2008.4735485

[24] Findorák R, Fröhlichová M, Legemza J. Potential of ilmenite sand application in the iorn ore materials agglomeration. Metalurgija. 2014;**53**:9-12

[25] Li Y, Li H, Wang H, Qing S, Hu J, Hou Y, et al. Smelting potential of HIsmelt technology for high-phosphorus iron ore and ilmenite. In: International Conference on Computer Distributed Control and Intelligent Environmental Monitoring, 19-20 February 2011. Changsha, China: IEEE; 2011. pp. 1283-1286. DOI: 10.1109/CDCIEM.2011.353

[26] Hou YL, Quig S, Wang H, Shi Z, Li H-b. Mixture of ilmenite and high phosphorus iron ore smelted by oxygen enriched top-blown smelting reduction. Journal of Central South University. 2012;**19**:2760-2767

[27] Chun L, Bin L, HaiYu W. Preparation of synthetic rutile by hydrochloric acid leaching of mechanically activated Panzhihua ilmenite. Hydrometallurgy. 2008;**91**:121-129

[28] Sasikumar C, Rao DS, Srikanth S, Mukhopadhyay NK, Mehrotra SP. Dissolution studies of mechanically activated Manavalakurichi ilmenite with HCl and H2SO4. Hydrometallurgy. 2007;**88**:154-169

[29] Vijay P, Venugopalan R, Sathiyamoorthy D. Preoxidation and hydrogen reduction of ilmenite in a fluidized bed reactor. Metallurgical and Materials Transactions B. 1996;**27**(B):731

[30] Reeves JW. Wilmington, Del. Method of ilmenite reduction. Filed on

November 9, 1972, Application No. 305 082, Granted on November 24, 1974, Patent No. US 3,850,615

[31] Paswan D, Venugopalan T, Malathi M. Journal of Metallurgy and Materials Science. Oct-Dec 2017;**59**(4): 163-171

[32] Gupta RC. Theory and Laboratory Experiments in Ferrous Metallurgy. New Delhi: PHI Learning Pvt, Ltd.; 2010. p. 213

Chapter 6

# Recent Trends in the Technologies of the Direct Reduction and Smelting Process of Iron Ore/Iron Oxide in the Extraction of Iron and Steelmaking

*Joseph Ekhebume Ogbezode, Olusegun Olufemi Ajide, Oluleke Olugbemiga Oluwole and Olusoji Ofi*

## Abstract

The blast furnace and direct reduction processes have been the major iron production routes for various iron ores (i.e. goethite, hematite, magnetite, maghemite, siderite, etc.) in the past few decades, but the challenges of maintaining the iron and steel-making processes are enormous. The challenges, such as cumbersome production routes, scarcity of metallurgical coke, high energy demands, and high cost of production, cannot be overemphasized. This study provides a systematic overview of the different ironmaking routes, their operational limitations and proper sound future perspectives to mitigate the challenges involve based on iron production demands in the modern-day metallurgical process. Subsequently, strategic ways toward improving the production efficiency and product quality of metallic iron produced in the recent iron processing routes were suggested. The study reiterated that the non-contact direct reduction and reduction-smelting routes are the faster ironmaking and steelmaking processes that can utilize alternative energy sources efficiently with little or no carbon deposition. Both processes also have promising features based on their requirements in terms of fewer energy demands, time-saving, cost-effectiveness, and operational efficiency. Thus, in today's iron and steelmaking processes, non-contact direct reduction and reduction-smelting processes remain viable alternative iron production routes.

**Keywords:** iron ore, blast furnace, direct reduction, smelting process, ironmaking and steelmaking

## 1. Introduction

### 1.1 Study background

The challenges associated with iron ore processing have resulted in rigorous research of new processing techniques in the iron and steelmaking process, considering their diverse applications in the construction and manufacturing industries. Iron ore consists of various elements within its metal matrix with the presence of silicate and aluminate as the gangue materials. A trace amount of manganese, titanium, nickel, etc. can be found within their microstructure. Iron ore processing has its root in hematite ($\alpha$-$Fe_2O_3$), maghemite ($\gamma$-$Fe_2O_3$), magnetite ($Fe_3O_4$), goethite ($\alpha$-FeOOH), siderite ($Fe_2CO_3$), etc. which can be prepared as nuggets for major feeds in the blast furnace as well as other conventional iron-making processes [1]. The focus of most research from the works of literature revolves around the production of iron metals through recycling or improving upon the existing method of production in the iron and steelmaking industry [2–4]. These methods include the Blast furnace/Blast Oxygen Furnace [5, 6], Direct Reduction Method [7], Indirect Reduction Method [8, 9], and Reducer-Smelter Process [10–12]. Thus, it is, therefore crucial to know that this processing technique has proven to have produced quality iron but not without a high cost of production, increased environmental pollution, and cumbersome production process [13, 14]. Therefore, an in-depth approach to producing metallic iron that is cost-effective, low risk of environmental pollution, and easy production route becomes imperative [15–17]. Thus, this paper attempts to investigate the challenges militating against the various iron extraction routes in the modern-day iron and steel-making process, Special attention was paid to the analysis of the metallurgical operation behind the conventional iron and steelmaking process, while an extensive overview of the merits, advantages, limitations and future perspective on improvement strategies for the conventional iron and steel production routes were also substantiated.

Also, a new iron extraction technique known as the non-contact direct reduction and smelting process was introduced as a decarburization mechanism and preventive approach to mitigate the direct interaction of harmful elements such as sulfur and phosphorus in conventional iron and steelmaking processes. The processing parameter for the production of direct reduced iron (DRI) in the conventional ironmaking process was also meticulously investigated to mitigate the effect of carbide formation across the intra-granular surface of the DRI which is likely responsible for environmental pollution due to high carbon deposit from the iron and steel production in modern-day metallurgical processes. The carbon-bearing substance and non-carbonaceous materials utilized for this study are non-coking coal, charcoal, biomass, methane ($CH_4$), hydrogen ($H_2$), and carbon monoxide (CO).

## 2. The iron ore/iron oxide reduction technique

Iron ore or iron oxide reduction is the process of removing oxides from ferrous materials. Some of these ferrous materials include hematite, magnetite, siderite, goethite, etc. Several processes have been postulated in the works of literature. Among the reduction methods of iron oxides in ironmaking processes are direct reduction, blast furnace/blast oxygen furnace, and smelting-reduction process. Reduction of iron oxide involves the upgrade of iron ore in pellets, lumps, or sintered form under a highly thermal-charged atmosphere in the presence of carbonaceous materials [18, 19], agric waste [20], or reducer gases [21].

### 2.1 Direct reduction process

Direct Reduced Iron (DRI) is the product of the reduction of iron ores (lumps, pellets, or fines) in solid-state by gaseous or carbonaceous substances [22, 23]. The reduction reaction temperature range generally occurs between 450 to 1300°C. The DR technique is a broad group of processes based on different feedstocks, furnaces, and reducing agents. Its common operation principle entails the removal of oxygen (reduction) from iron ores in the solid state, unlike the blast furnace method that comprises both solid and liquid state iron oxide reduction. Natural gas (in some cases coal) is used as a reducing agent in this method of iron extraction. In the year 2000, about 92.6% of DRI was based on natural gas processed in shaft furnaces, retorts, and fluidized bed reactors [24]. The metallization rate of the end product ranges from 85–95% (often even higher). The product from this process is called DRI or sponge iron or hot briquette iron (HBI), which is a highly metalized solid that still contains gangue minerals, and hence needs further purification [25].

The direct reduction method was adopted by Delta Steel Plant in Nigeria, while they were still in operation [26]. Most steel plants across the globe also adopt this method, by making use of ferrous scraps. **Figure 1** provides a layout of the direct reduction of the process, while **Figure 2** provides the general schematic flow for performing the direct reduction process of iron ore/oxide in a carbonaceous atmosphere from lump charcoal/non-coking coal with the view to further give the ironmaking process a more robust perspective during the production of pure iron.

In this process, iron is extracted from its ore at a temperature below the melting points of the materials involved. The method is used mostly in special circumstances, often linked to cheap supplies of natural gas. Gojic et al. [28] have identified different challenges associated with the direct reduction process of which MIDREX (named after its developer, a division of the Midland-Ross Corporation) is the world's leading technology.

Thus, the common challenge which is peculiar to the direct reduction process of iron ore/oxide is the need for a regular supply of natural gas (in the case of gaseous reductants), which may be distorted by the incessant attack on gas pipelines in some parts of the world. Gangue remains in the sponge-like product, known as DRI), which must be removed in subsequent ironmaking and steelmaking process, hence more cost is accrued. Only high-grade ores and pellets made from super-concentrated iron ore (above 66% iron) are suitable for the DR iron-making process, therefore beneficiation of iron ore samples is needed and this may also increase the cost of production.

## 2.2 Indirect reduction process

This involves the production of iron using other sources of reductant substances such as carbon II oxide, methane, hydrogen gas, etc. it comprises of solid-gaseous reduction process of iron ore other than coke. It is majorly the strategy for producing

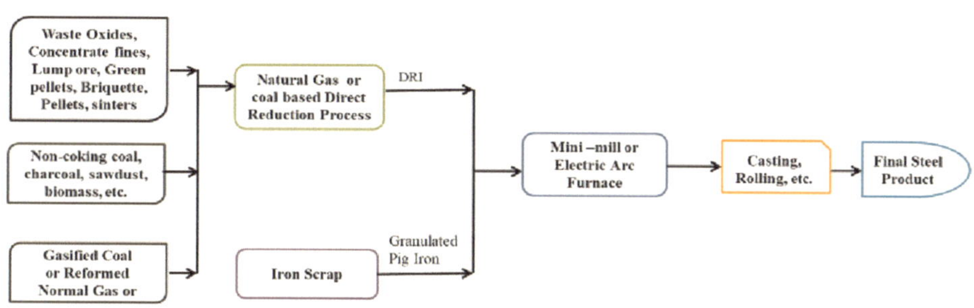


**Figure 1.**
*Schematics of iron and steel production routes by direct reduction method [27].*

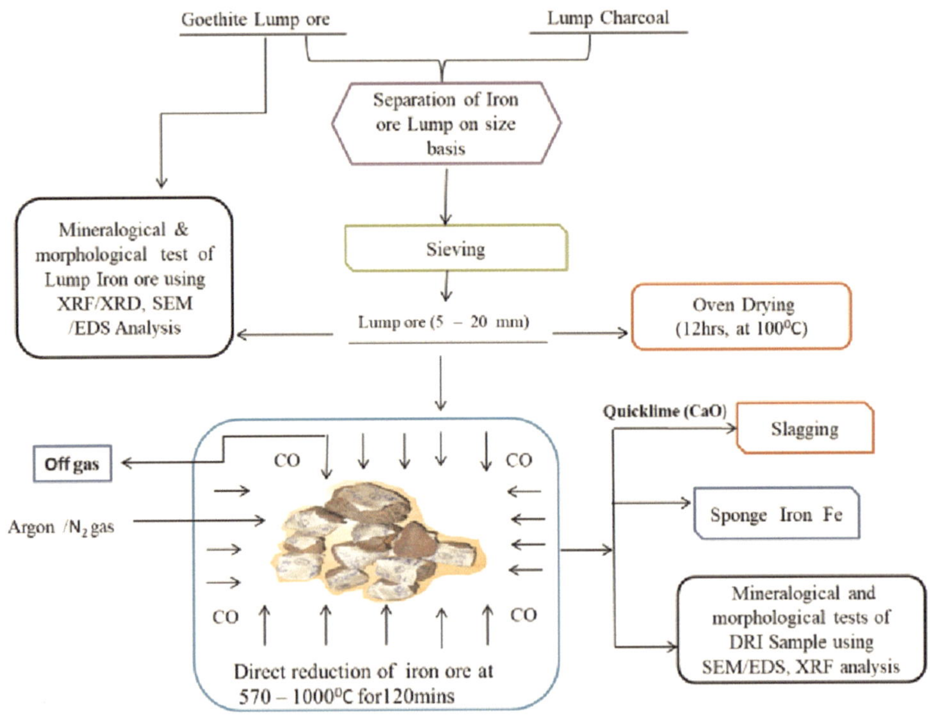


**Figure 2.**
*Schematics of an experiment flow chart of direct reduction of iron ore/oxide.*

metallic iron as an alternative route to the blast furnace or blast oxygen furnace and electric arc furnaces. This also involves the use of pure carbon monoxide (CO), hydrogen ($H_2$), methane ($CH_4$) gas, etc. as the major reductants. For the major gaseous reductants (i.e. CO, $H_2$, and $CH_4$) in a DR process, the following three reduction reactions can be written as shown in **Table 1**.

The indirect reduction approach can also be expensive as it is unadvisable for usage in iron and steelmaking processes at an industrial scale. Majorly, the indirect reduction method can only be sustainable when carried out on a laboratory scale. Though, iron nugget produced from this approach tends to be gangue free and often free of radical elements such as sulfur and phosphorus which may interfere with the chemical activities within the microstructure phase of the heat-treated iron ore. Furthermore, the activity of noble gases such as argon is employed in the indirect iron reduction process to mitigate the interference of other external gases from taking part in the reduction process. The reduction reaction between solid iron ore and carbon monoxide entails the removal of oxygen from higher oxide of iron under strict experimental conditions. The reaction can be a solid–gas or liquid–gas reaction where the release of carbondioxide gas and lesser oxides of iron are acquired as the product of the reduction process.

### 2.3 Non-contact direct reduction (NDR) process

The Non-contact Direct Reduction (NDR) process is a novel approach aimed to mitigate the effect of excess carbon deposit on DRI which is most common in the iron product produced via other conventional iron-making processes. This approach tends

| Iron Ore/ IronOxide | Reduction with CO | Reduction with $H_2$ | Reduction with $CH_4$ |
|---|---|---|---|
| Hematite | $3Fe_2O_3 + CO \rightarrow 2Fe_3O_4 + CO_2$ | $3Fe_2O_3 + H_2 \rightarrow 2Fe_3O_4 + H_2O$ | $12Fe_2O_3 + CH_4 \rightarrow 2Fe_3O_4 + CO_2 + 2H_2O$ |
| Magnetite | $Fe_3O_4 + CO \rightarrow 3FeO + CO_2$ | $Fe_3O_4 + H_2 \rightarrow 3FeO + H_2O$ | $3Fe_3O_4 + CH_4 \rightarrow 3Fe + CO_2 + 2H_2O$ |
| Wustite | $FeO + CO \rightarrow Fe + CO_2$ | $FeO + H_2 \rightarrow Fe + H_2O$ | $Fe + CH_4 \rightarrow Fe + CO_2 + H_2O$ |

**Table 1.**
*Iron ore/oxide reduction pattern for common reducing agents in gaseous forms.*

to also bypass the idea of pelletization, communition, and hot briquette preparation of iron ore-reductants composites which have shown to have high economic implications, are energy-demanding, time-consuming, etc. Ogbezode et al. [29] introduce the concept of the ironmaking technique the NDR method by investigating the reduction behavior of goethite ore using carbon monoxide gas from the wood charcoal atmosphere. The result revealed a DRI with a metallization degree of over 93% was achieved at an overall reduction temperature of 1000°C. Ogbezode et al. [30] also performed an NDR process on selected goethite-hematite ore under isothermal conditions using $CO/CO_2$ atmosphere from wood charcoal. The kinetics of the reaction and rate-controlling resistances of the process was monitored at different roasting temperatures and reaction rate. The study affirmed that the presence of an ash layer deposit from the charcoal which is transported by gas penetration from the burned charcoal forms an interstitial layer on the DRI and was responsible to have slowed down the rate of reaction between iron ore and the reducing gas. The advantage of this direct reduction method is that the DRI produced via the process does not accommodate any carbon deposits and sulfur within the metallic iron phase microstructure.

The limitation of the non-contact reduction method of iron ore is the presence of gangue materials that may be present as impurities within the metal matrix of the DRI. Thus, for effective and optimum performance, the DRI produced by the non-contact direct reduction method should be smelted to eliminate gangue minerals from its metal structure using quicklime (CaO) or limestone ($CaCO_3$) at a smelting temperature above 1400°C. A schematic view of the NDR process is depicted in **Figure 3**.

## 2.4 Reduction-smelting process of iron ore/oxide using solid reductants

An alternative ironmaking and steelmaking process that utilizes other sources of energy instead of the conventional metallurgical coke used in blast furnace operations can be referred to as a smelting-reduction process. It is a metallurgical operation with a dual-compartment approach where the pre-reduction and post-reduction process takes place. The pre-reduction chambers utilize iron ore in pellets, lumps, or sintered form to a metallization level of over 90%, in the $CO/CO_2$ atmosphere from any carbon-bearing material at a reduction temperature of between 570–1300°C. Then, the reduced iron ore is then charged into the post-reduction chamber for the smelting process at a temperature above 1400°C. This operation is aimed to actualize a slagging process through the introduction of limestone ($CaCO_3$) or quicklime (CaO). The pure metallic iron condenses, while the gangues are fluxed out as aluminate and silicate

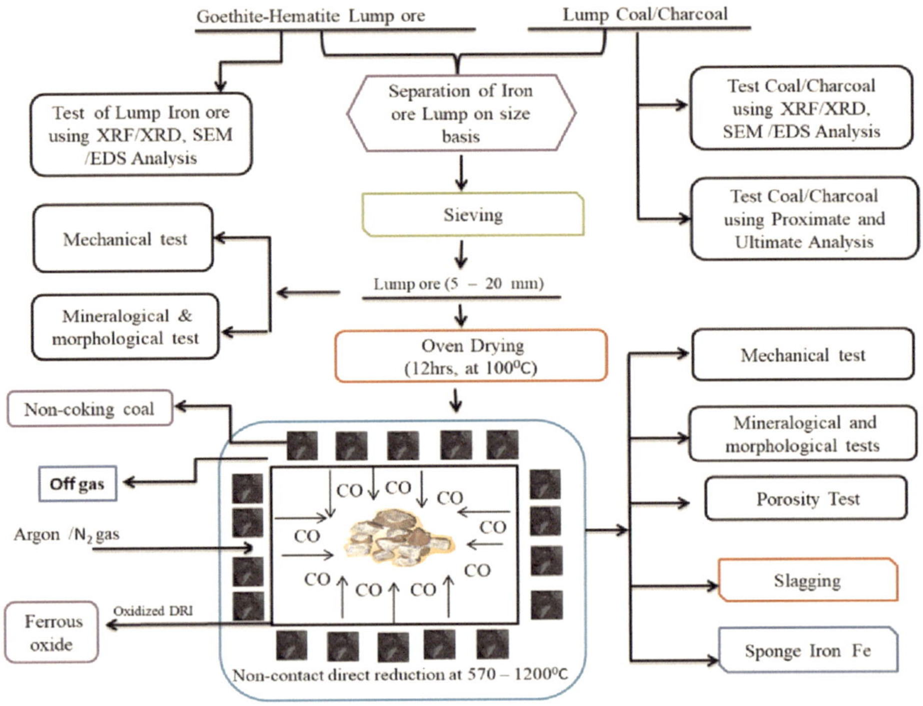


**Figure 3.**
*A schematic view of the non-contact direct reduction process of iron ore/oxide.*

compounds known as slags. Liquid iron or pig iron is obtained as the final product which can then be subject to the casting and rolling process to obtain a final product as steel. **Figure 4** tries to depict a modification of the Anameric et al. [27] approach for the iron reduction-smelting process, the modification was necessary to mitigate the possible limitations that may affect the process in the long run. For instance, this can be envisaged by introducing a mini-reduction route within the ambiance of the post-reduction route through the use of hot reduction gas by de-oxidation of carbon dioxide gas released from the post-reduction vessel as shown in **Figure 5**.

In a typical reduction-smelting process, the iron ore will first undergo a solid-state reduction in the pre-reduction unit. The resulting product is then smelted and further reduced in the smelting-reduction vessel where the coal gasification occurs, thus delivering heat and carbon-monoxide-rich hot gas. Coal gasification takes place due to the reaction between oxygen and iron ore in a liquid state by the thermal combustion process. The heat is used to smelt the iron, while the hot gas is transported to the pre-reduction unit to reduce the iron oxides that enter the chambers. Reduced iron oxides (similar to DRI) are in turn transported to the smelting-reduction vessel for the final reduction and smelting process. The carbon-monoxide-rich gas generated in the post-combustion chamber can be further oxidized to generate additional heat for the complete reduction and smelting of the pre-reduced DRI.

The direct reduction process takes advantage of the availability of cheap natural gas and/or hydroelectric power, as well as access to suitable iron ores, and agglomerates [32], whereas the reducer-smelter process takes advantage of the accessibility of non-coking coals or reducer gas with low energy requirements, low operating costs, the ability to respond flexibly to variations in production rate, and low environmental

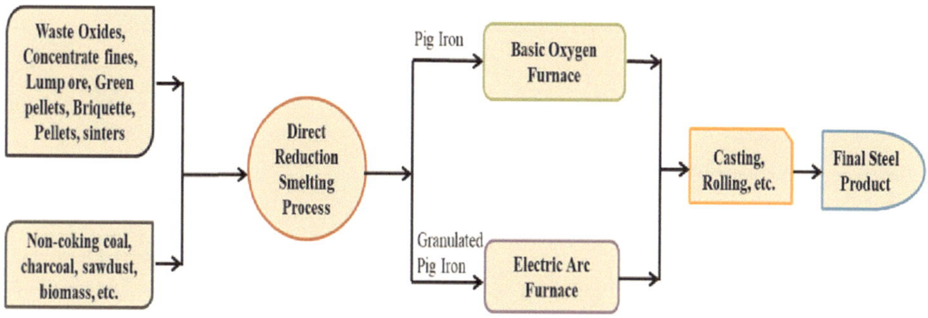


**Figure 4.**
*Schematics of iron and steel production routes by direct reduction-smelting process [31].*

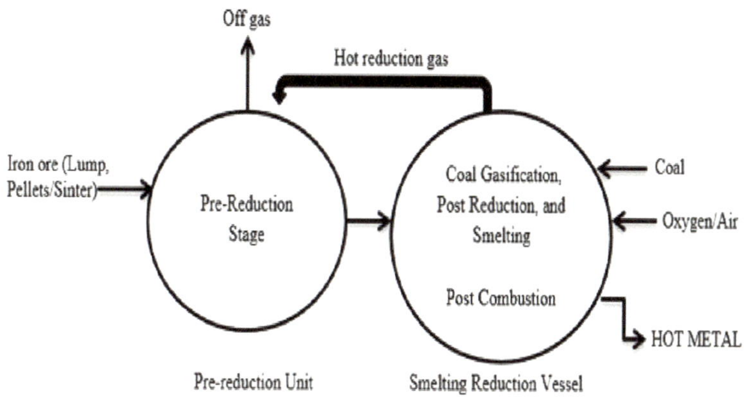


**Figure 5.**
*Layout of a typical smelting reduction process [31].*

loads, as there will be no coke oven gas leaks. Because the approach does not necessitate the removal of undesirable materials from the iron ore sample, this step can be bypassed during the iron extraction process [33]. Thus, the numerous challenges that accompany the conventional iron ore processing routes cannot be overemphasized. This limitation includes the interference of foreign substances on the metallic, low capital investment, high cost of production, high energy consumption, scarcity of metallurgical coke, etc.

To address these limitations, the smelting-reduction technology must be complemented with a decarburization mechanism inside the furnace, and the use of non-coking coal, charcoal, biomass, etc. as alternative energy sources in the ironmaking routes (see **Figure 6**). Thus, the advantage of the smelting-reduction ironmaking technology is also due to its operational feasibility, cost competitiveness, and ability to utilize all kinds of iron ore materials in their natural forms (pellets, lump, sinter, etc.). The process can also utilize other reductants (solid, liquid, gas, etc.) as alternative energy sources instead of conventional metallurgical coke [34].

### 2.5 Reduction-smelting process of iron ore/oxide using gaseous reductants

The quest for the production of iron and steel under clean energy evolution cannot be overemphasized. The negative impact of carbon emissions released from the

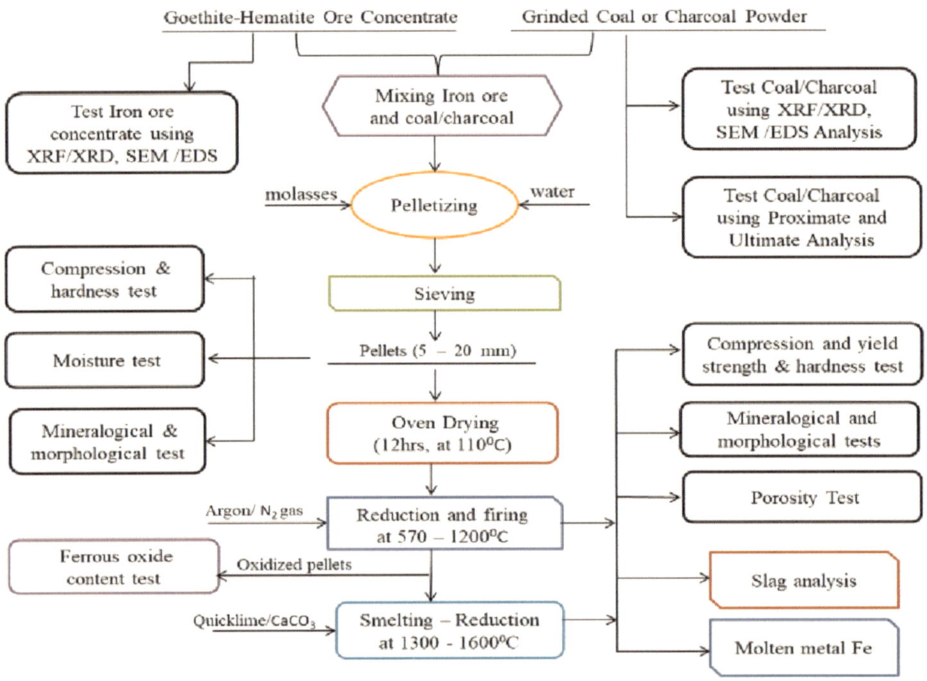


**Figure 6.**
*Schematic flow chart of the reduction-smelting process.*

metallurgical industry on a global scale has given the need to use less harmful reducing gases as a replacement for fossil fuel for the decarburization of the iron and steelmaking process. Heidari et al. [35], explain the need for a green footprint in the metallurgical industry through the use of pure iron oxide and reducer gases. Certain variables such as reduction gas flow rate, $H_2/CO$ ratio, diluents concentration (Ar, $N_2$, and He), gas utility, and pressure are among the parameters that foster the production of pure iron using gaseous reductants.

The use of hydrogen unlike CO gas in the metallurgical process has been attempted by many researchers [36, 37], in the quest to produce pure metallic iron in present-day ironmaking technology. The kinetics of the reduction of iron oxide in $H_2$ gas was investigated by Gonoring et al. [38] as a strategy to mitigate the energy consumption rate in conventional metallurgical processes. The study reiterates the need for $H_2$ gas used as an excellent reducing agent in the alternative metallurgical process of iron and steel. Hou et al. [39] conducted a general study on iron oxide reduction kinetics in a hydrogen gas-infested atmosphere under isothermal conditions in a micro-packed bed, influence of diffusion and reduction reaction mechanisms are the major reaction kinetics stages that were enumerated. The magnetite to wustite stage of the reduction process was eliminated leaving behind the hematite to magnetite stage and wustite to the metallic iron stage since the internal and external gas diffusion causes an increase in $H_2$ gas into the inner core of the iron oxide particle provided the reduction process is done at reduction temperature above 570°C. Thus, the intrinsic reduction kinetics of iron oxide can be done in an $H_2$ atmosphere using a micro-packed bed giving reaction kinetics free of internal and external diffusion resistances. Du et al. [40], performed the intermediate reduction on hematite to magnetite ore under low thermal conditions using hydrogen gas. The study acknowledged that the utilization of hydrogen as a reduction rate of hematite to magnetite increases the overall theoretical iron recovery as the rate of hematite conversion increases. Hydrogen remains an efficient reducing agent which has a tremendous effect on iron oxide reduction depending on the reducing flow rate and thermal condition being utilized. Vogl et al. [41] conducted an extensive assessment of the direct reduction of iron oxide void of carbonaceous substances. Ma et al. [42] investigated the reduction behavior of iron oxide pellets in a combined atmosphere of CO, $H_2$, and $N_2$ at a reduction temperature of 900°C. A schematic illustration of the smelting-reduction process of iron oxide using gaseous reductant is depicted in **Figure 7**.

The study substantiates the thermodynamic response of the iron oxides at different reducer gas flow rates under controlled atmospheric conditions. These works of literature do establish the fact that the assessment outlook of the direct reduction process of iron oxide in conventional steelmaking technology does ride on the premise that the use of gaseous reducing agent iron and steel technology has undoubtedly triggered a metallurgical advancement in iron and steelmaking process with undeniable impact on low economic and environmental implications and optimum overall technological development in the metallurgical industry. Naseri-Seftejanim et al. [43], examined the use of hydrogen in the reduction of hematite ore by the Hydrogen Plasma Smelting Reduction (HPSR) process. The ironmaking technology approach entails the use of hydrogen plasma in the development of a carbon-monoxide-free metallurgical process. In the HPSR process, influencing parameters such as hydrogen utilization degree, iron oxide reduction degree, weight loss, rate of reduction, etc. can be assessed using appropriate experimental conditions. Consequently, the thermodynamic reduction of liquid iron was performed by Naseri-Seftejani et al. [44] using hydrogen plasma. HPSR entails the use of $H_2$ in a plasma state for the reduction of iron ore. The study

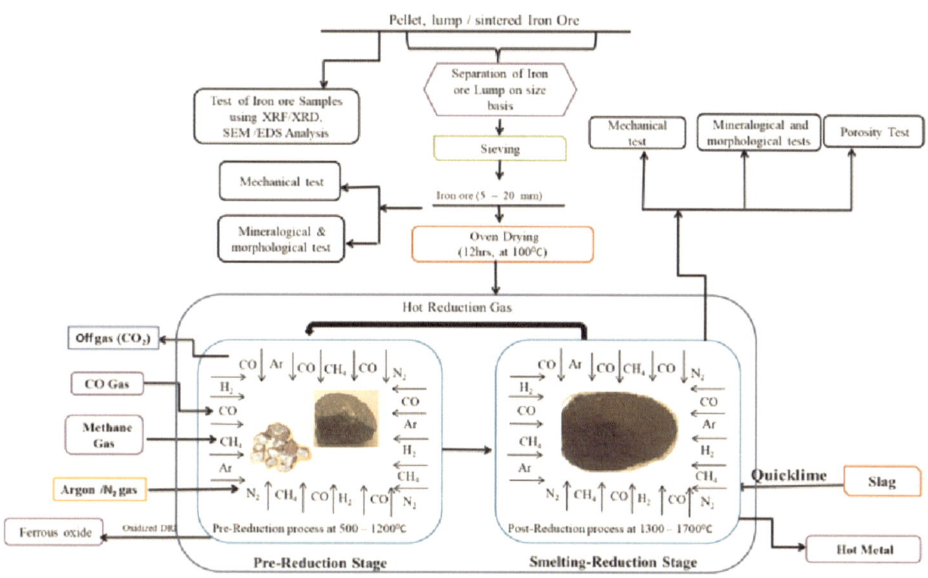


**Figure 7.**
*Schematic flow chart of the reduction-smelting process of iron ore using gaseous reductants.*

presents the hematite reduction reaction sequence, hydrogen ionization degree, iron ore particle reaction rate, and reduction temperature as major parameters which affect the overall quality of the molten iron and hydrogen solubility in slag formation produced under severe thermal reduction of the iron ore sample in hydrogen plasma as the most influential of all the reduction parameters. The slag formation from an iron oxide reduced in the HPSR method was investigated by Naseri-Seftejani et al. [45], based on the quest to eliminate the challenges portrayed by $CO_2$ emission in convectional iron and steelmaking technology. The study makes a presentation on reduction degree, operational hydrogen utilization, produced iron, and reduction operational time remains the vital components in slag formation in the HPSR process of iron and steelmaking process under a hydrogenated atmosphere. Thus, the use of the HPSR process in iron and steelmaking technology has proven to be more effective in $CO_2$ elimination. The use of HPSR also produces liquid slag with lower iron oxide content compared to other convectional iron-making processes.

On a general note, the Smelting-Reduction (SR) process of iron oxide can utilize several carbonaceous and non-carbonaceous substances as reducing agents. These reducing agents include metallurgical coke, non-coking coals, charcoal, agric wastes, biomass sawdust, reducer gases ($H_2$, $CO/CO_2$, $CH_4$), etc.

### 2.6 Iron ore reduction-smelting process using non-coking coals/charcoal as reductants

The use of non-coking coal as an alternative energy source for sponge iron production instead of the conventional metallurgical coke cannot be overemphasized due to the problematic nature that is common to coke making process. The non-coking coal has the advantage of not requiring preprocessing technique as it can be charged into the reactor during sponge iron production. Another interesting fact about non-coking coal remains the fact that it has wider applications in many iron processing reactors or furnaces, compared to that metallurgical coke whose traditional strategy for application is found in the blast furnace. Iron production processes such as FASTMET [46], HISMELT [47], MIDREX [34], COREX [48], Microwave ovens [49], Rotary kiln [50], HIsarna [51], etc. have been used utilized in the works of literature as sponge iron ironmaking processes reactors. Moreover, most of the ironmaking processes have unlaying challenges which are currently been addressed by researchers in the field of materials and metallurgical engineering. The aforementioned iron-making processes which utilize non-coking coal as the alternation energy source are used to produce direct reduced iron DRI which occurs when producing iron metals below their melting point. The DRI also has high impurity content in form of gangue which is mostly locked within the clusters of the DRI crystal structure. Coal based DR process was done on iron ore/coal pellets under axisymmetric thermal conditions [23]. Furthermore, the coal-based reduction under the isothermal conditions on fayalite was conducted by Zhang et al. [52] using thermogravimetric analysis. Li et al. [53] studied the reaction kinetics by carbothermic reduction of fayalite under non-isothermal conditions. Ogbezode et al. [29] performed reduction runs on goethite iron ore under non-isothermal conditions using charcoal. Ogbezode et al. [30] also performed carbothermic reduction and investigated the kinetics behavior of goethite-hematite ore in the $CO/CO_2$ atmosphere. Man et al. [42] also examined changes in iron/ore pellet composite during the carbothermic DR process. Kumar et al.

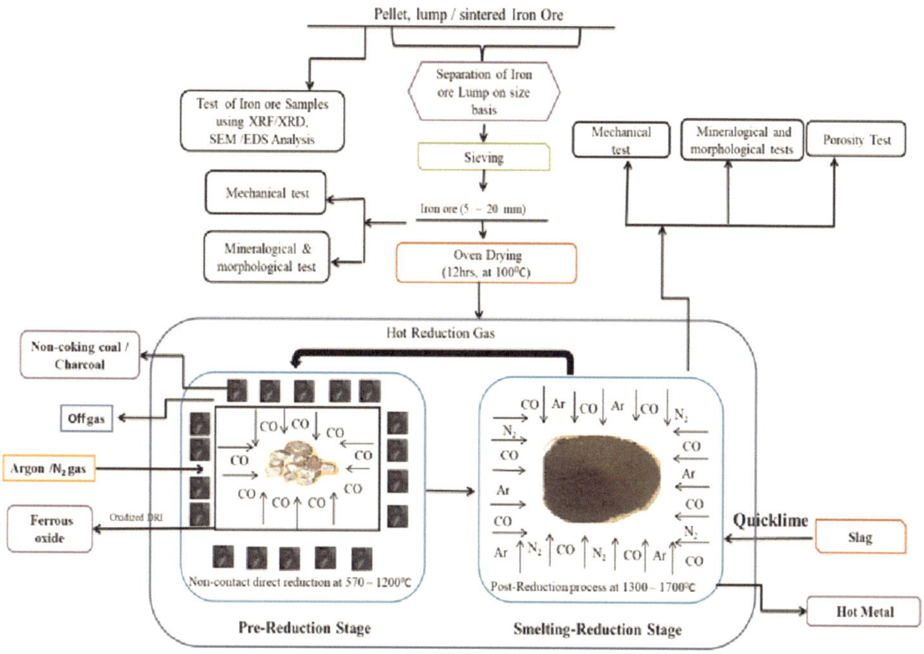


**Figure 8.**
*Schematic flow chart of the reduction-smelting process of iron ore using non-coking or charcoal as reductants.*

[54], investigated the characteristics of non-coking coal under non-isothermally reduc*ed* iron oxide. The study reiterates that increased devolatilization of coal at high temperature do have minimal effect on the reducibility rate of the iron ore. This implies that characteristic volatile matter associated with coal does have thermodynamic implications on the reduced iron oxide when heat treated [55]. Thus, the quality of DRI produced by carbothermic reduction of iron ore-coal composites depends on many factors ranging from quality of iron ore, availability of coal, increased thermal condition, reduction time, etc. A schematic illustration of the reduction-smelting process of iron ore/iron oxide samples using non-coking coal or charcoal as reductants is depicted in **Figure 8**.

Therefore, the availability of DRI using non-coking coal as a reducing agent is determined by the availability of iron ore with over 65% Fe quality and the availability of a cheap source of non-coking coal. Due to the operational limitation which accompanies the use of non-coking as the perfect alternative for metallurgical coke used in the blast furnace process, it is necessary to foresee the need to produce pure metallic iron in the direct reduction process. Going forward, the presence of high silicon and alumina content as gangue elements which pose serious limitations on the performance of direct reduced iron begs for the need for the metallurgical industry to source quicklime or limestone in its pure form in slagging operation. Pal et al. [56] performed a development slagging experiment by smelting operation of pellet iron ore using calcined lime instead of limestone and magnesium oxide in a carbonaceous atmosphere. Thus, the use of methane, hydrogen, and pure carbon monoxide gas became necessary in the quest to produce the needed feeds for the production of pure metallic iron.

## 3. Solid: solid reduction of iron ore by carbon

For solid carbon in a DR process, the following three reduction reactions can be written:

$$3Fe_2O_3 + C \rightarrow 2Fe_3O_4 + CO. \quad (1)$$

$$Fe_3O_4 + C \rightarrow 3FeO + CO. \quad (2)$$

$$FeO + C \rightarrow Fe + CO. \quad (3)$$

Only a negligible amount of reduction will occur by direct contact of carbon particles with iron oxide particles since such solid-solid reactions are very slow. The actual reduction process occurs through the intermediary of CO.

In the direct reduction process of iron ore where solid carbon is the reductant, self-reducing pellets, briquettes or composites entails iron ore embedded in coal or carbonaceous materials are typically used in blast furnace and direct reduction processes to mitigate the effect of carbon consumption for better productivity and performance. The limitation of the direct reduction process is the quality of coke or coal as they provide limited reaction conditions with low productivity.

### 3.1 Reduction of iron ore using coking coal

Iron oxide reduction by carbon is a significant metallurgical process. Theoretically, the reduction might entail either gaseous intermediates or direct reduction [57]. Carbothermic reduction of iron ore using coking coal has been carried out by several researchers and metallurgists using majorly direct reduction, magnetic separation, reduction-smelting, and thermo-mechanical processes. An extensive study on iron ore reduction in their composite aggregates was also reported in the existing literature. Hou et al. [39] investigated a reduction of lump hematite in a carbothermic atmosphere using a coal-based reduction and magnetic separation method. The hematite ore undergoes a pre-reduction process at a maximum temperature of 1000°C in a coal-infested atmosphere before the disintegrated direct reduction of the hematite ore particle is then subjected to magnetic separation to increase the metallic iron concentration of the DRI. Reduction of iron ore in lump or pelletized form in the presence of coking coal entails a solid-solid reaction mechanism where the CO released from the coke directly permeates the surface structure of the iron ore to obtain a metallic iron as a by-product. But the challenge with the use of coking coal in the ironmaking process can be traced to its scarcity and its undoubted tendency to produce DRI with a high carbon deposit. Studies have shown that ferrous metals produced under Ore/coal composite technique using high carbonaceous materials such as coal will give very hard but brittle and ferrous metals as a by-product [18]. Thus, such DRI might exhibit poor mechanical properties which include low malleability, ductility, and machinability tendencies. Man et al. [42] performed a thermogravimetric experiment on an iron/coal composite by direct reduction process at different reduction temperatures in a nitrogen atmosphere. The changes in the mass of the ore/coal composites reveal the extent mechanical degradation occurs based on the different mass loss, heating rate, and heating temperatures. Also, the permeability rate of carbon from the coking coal on the iron ore concentrate determines the quality of DRI expected to be produced from the reduction experiment. Also, an increase in carbon gas penetration around the interstitial surface of the banded iron/coal composite is strongly affected by increasing the heating temperature and heating rate [58]. This may in turn affects the overall performance of the direct reduction experiment process. Among the parameters that are majorly affected by increased carbonaceous activity with the iron/coal composite when the reduction temperature is increased are increased reduction degree, percentage metallic iron, high carbon deposit, and increased iron/coal composite degradation. In contrast, the scarcity of coking coal for the iron and steelmaking process has called for research concerns where metallurgists and material scientists have started looking into non-coking coal, charcoal, biogas, etc. Another viable alternative replacement for metallurgical coke is due to its complex method of preparation and negative environmental impact on the end-users.

### 3.2 Reduction of iron ore using non-coking coal/charcoal

For example, it is predicted that Nigeria has over 2 billion metric tons of coal reserves [59]. Although it once possessed a few small coal-fired power stations none of them are still in use today. The government plans to use coal to generate 30% of the nation's electricity, according to repeated statements from the mining ministry. Additionally, some observers think non-coking coal could help Nigeria mitigate its energy crisis [60]. For instance, in some coal-rich states in the north-central region of Nigeria, a 1200-megawatt (MW) coal-fired power station was envisioned [61]. The project has been frequently put on hold due to financial issues [62]. Later, two more coal plants were planned, although no plans have been made public since the government announced its intention to build them in 2011 [63]. Thus, the socio-economic and manufacturing shortcomings ravaging the usability and accessibility of coking coal in the iron and steel industry have led to the need to seek other reductants as alternative energy sources (i.e. non coking coal, charcoal, biogas, etc.) which can be used in the production of DRI in convectional iron and steelmaking process. Several works of literature have attempted to research the feasibility and viability of these reducing agents using different ironmaking production routes as tremendous processes have been reported about some of the DRI produced when compared to those of the DRI produced in the metallurgical industry [64, 65]. These have shown to have similar and even better quality in terms of their physical, metallurgical, and mineralogical properties.

Although lump iron ore is less expensive than the pelletized version, they have historically been limitations on its usage in blast furnaces due to long-standing worries about its deterioration and high-temperature characteristics. As a result of high-grade iron ore depletion in supplies, gangue minerals are also becoming more prevalent in iron ore fines, which have caused a significant change in their sinter chemistry, particularly in terms of alumina content and basicity.

Lu et al. [66] performed a direct reduction and magnetic separation process on lump hematite ore using lignite as a reductant. Lignite is low-grade non-coking coal of low ash and higher carbon content than bitumen. The reduction roasting was done at 1200°C with 35% lignite on low-graded lump hematite ore with a particle size range of (20–50 mm) at a reduction time of (60–300 mins), the reduced iron ore was ground into a two-stage concentrate (20/30 mins) of 90% theoretical Iron (TFe) and 89.3% iron recovery. The effect of reduction time, roasting temperature, particle dosage, and grindability on the metallization degree, iron recovery, and microstructure phase change on the reduced hematite ore were investigated. Ogbezode et al. [29] examined the kinetic reduction behavior of goethite-hematite ore using wood charcoal in an activated carbon furnace. The DRI produced was achieved through the non-contact reduction of goethite-hematite ore acquired as run-off mines from selected ore depots. The study aims to investigate the behavioral pattern of iron ore in the $CO/CO_2$ atmosphere from wood charcoal. The kinetic model employed was the shrinking core model. The model attests to the fact that goethite ore can be reduced under the $CO/CO_2$ atmosphere where diffusion through the ash layer is the rate-controlling resistance of the reaction kinetic model. The microstructure phase showed a trace amount of carbon deposits on the direct reduced iron compared to using coking coal as carbon-bearing materials regardless of reduction temperature and reaction residence time [30]. From existing literature, the use of non-coking or charcoal as an alternative energy source instead of metallurgical coke in the production of direct reduced iron is targeted toward reducing energy consumption, decreasing carbon deposition, and eliminating complex procedures that accompany the conventional iron sintering processes [32, 67]. Also, the high energy demand and complex procedures involved in the pelletization process of iron ore, coupled with the scarcity of coking coal have made metallurgical scientists begin research into the direct use of iron ore in lump form and the use of non-coking coal or charcoal as reductant [20]. Production of DRI using iron ore in lump form may seem to be a viable alternative for already beneficiated iron ore provided the base material is of high-grade quality. Furthermore, the use of low-grade iron ore could better be encouraged provided a more efficient coal-based direct reduction DR process is employed. The treatment of low-grade iron ore material may hold promise for the coal-based DR process in the future. Therefore, solid carbon or gaseous carbon monoxide CO and hydrogen gas $H_2$ are needed as reducing agents in commercial techniques used in the removal of oxygen from iron ore. The latter can also be obtained naturally through carbon gasification [68]. Thus, it is advisable to take into account the usage of non-coking coal, first directly and then as a gasifying agent to produce the reductant components of CO and $H_2$ gases. Agrawal et al. [49] performed carbothermal reduction on low-grade magnetite ore in a microwave at an irradiation temperature above 600°C using coal and charcoal. There are two different types of composites were prepared which are made of iron ore fines and charcoal, respectively. Magnetite reduction with coal was discovered to be lean, resulting in poorer metallization; yet, the concentrate's increased iron-oxide content and also increases the iron extraction in terms of performance as compared to charcoal [69]. Regardless of thermal conditions, iron ore is reduced under the same heating rate, and coal has a lower reduction rate compared to charcoal.

The optimum performance of the reduction roasting of low-graded iron ore using non-coking coal should better reveal high theoretical percentage iron (TFe) of good magnetic concentrate which can produce metallic iron with higher purity, which means that a high-temperature reduction process should be encouraged in such cases. Thus, to effectively perform a coal-based reduction on low-grade iron ore using non-coking coal, a high reduction temperature, increased roasting time, and coal with high carbon content, low moisture, and low ash content should be used in other to achieve a sustainable reduction process. The carbon-iron ore dosage must be understood to ascertain the reduction behavioral pattern and phase transformation of the various gangue components present within the metal matrix of the coal-based reduced iron. The non-availability of solid carbon-bearing materials has triggered the need for metallurgical experts to begin to harness a more efficient approach to the production of molten iron in its pure form due to the presence of unwanted substances such as sulfur, phosphorus, lead, etc. that might have accompanied such DRI products. This concern among others does initiate the need to commence further research into the gas-solid reduction of iron ore or iron oxides. Consequently, further studies into the indirect reduction process of iron ore are majorly carried out on a laboratory scale due to their high production cost, purity of DRI product, environmental friendliness, and less hazardous process where its major limitation remains its high cost of production at an industrial scale.

## 4. Gaseous-solid reduction in iron ore processing

Many researchers have postulated different ways of producing quality iron using gaseous reductants such as carbonmonoxide [14], Hydrogen [70], Methane [71], biogas [72], etc. This is a method of producing direct reduced iron DRI using gaseous reductants as carbon-bearing material. To comprehend the use of gaseous carbonaceous materials as iron oxide reductants, the different phase changes within the macro and microstructure of the processed iron oxides must be known. A schematic illustration of the ironmaking process by direct reduction method using gaseous reductants is depicted in **Figure 9**. The effect of thermal decomposition, reduction time, and reaction kinetics surrounding the type of reduction mechanism and metallurgical processes used in the production of such metallic iron must be ascertained.

### 4.1 Reduction of iron ore by hydrogen and carbon-monoxide gas

Due to the adverse effect of carbon-infested gases on our immediate environment, the manufacturing and metallurgical industries produce crude steel of over 1.7 billion tons, while almost the total production of up to 75.1% is carried out

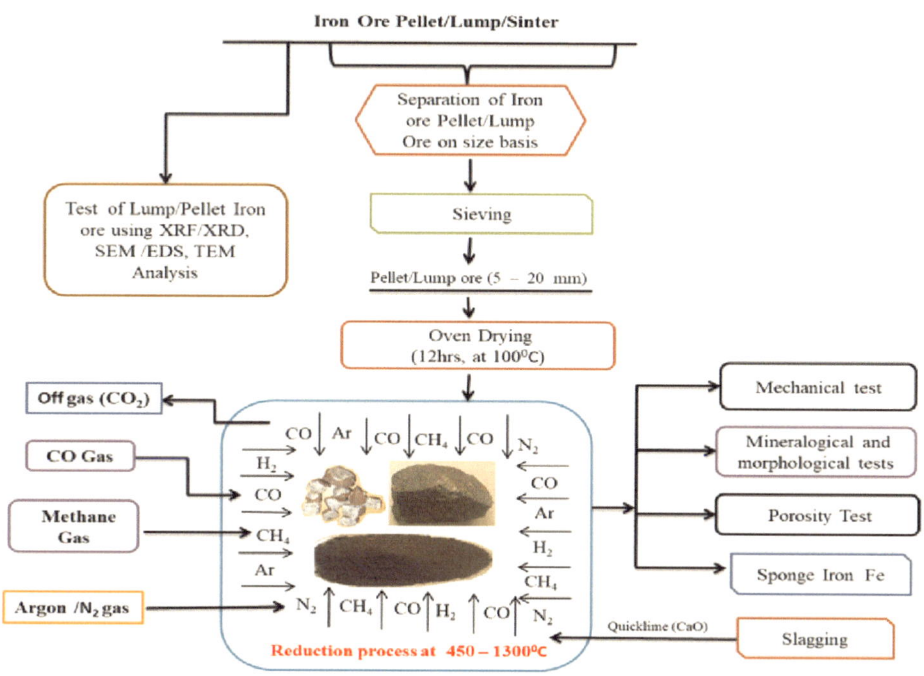


**Figure 9.**
*Schematic flow chart of the reduction process of iron ore using gaseous reductants.*

using coke-injected blast furnaces [73]. It is no longer news that coking coals can serve as an excellent reducing agent and high energy source for iron and steel production. But the under-utilized coke is produced as carbon monoxide gas which has an adverse climatic effect on our immediate environment [8]. The carbon monoxide gas may also lead to ozone depletion, the greenhouse effect, and global warming at large. These and many more sponsored the initiative for the use of other sources of carbon-bearing substances and gases as alternative energy sources instead of coke for iron ore reduction. This implies that the use of hydrogen, methane, fossil fuels and biomass is fast replacing coke, especially in metallurgical operations. Lately, hydrogen reduction has proven to be a viable alternative in the ironmaking process compared to the conventional carbothermic reduction processes. Li et al. [70] conducted a direct reduction process of iron ore using hydrogen gas ($H_2$). The hydrogen was formed by the catalytic breakdown of ammonia and methane as a way to mitigate $CO_2$ emissions. The by-product hydrogen gas is passed into the fluidized bed reactor containing the sets of cohesive hematite ore particles at a reduction temperature of 500°C and above. Several works of literature [37, 40, 41] have investigated the energetic and environmental essence of the hydrogen-based iron production process. The literature further reiterated that $H_2$_based direct reduction of iron ore possesses a carbon-free iron product with a prospective reduction of harmful gas emissions regardless of the kind of reduction or smelting reactor used [74]. Since the formation of iron-carbide is majorly associated with iron oxide reduced with carbon-monoxide, the use of hydrogen gas as the alternative reductant in the ironmaking process cannot be overemphasized. Another attribute of $H_2$-based direct reduction of iron ore is the physical characteristics and reduction kinetics mechanism associated with such an iron reduction process. These characteristics are reduction temperature, reaction control mechanism, particle size, reduction gas penetration rate, mineral phases, and morphological dependence [52, 75]. Consequently, the behavioral changes in the mineralogical phases of iron ore sinter due to the effect of hydrogen gas concentration were investigated by Xing et al. [76]. The study of mineral phases on iron ore sinter reduction in hydrogen atmosphere has proven to mitigate the effect of $CO_2$ emission in the ironmaking process. Also, whisker growth in direct reduced iron can be attributed to the presence of sulfur present in the reductant coals. As the presence of sulfur in the carbonaceous atmosphere during the ironmaking process possesses the tendency to initiate catastrophic swelling in the iron ore-carbon monoxide atmosphere which may produce a large number of iron whiskers. The presence of sulfur in the $CO/CO_2$-infested iron ore reduction atmosphere is triggered by the nucleation effect of sulfide inhibits gases, resulting in whisker growth. To eliminate the occurrence of whiskers on iron oxide surfaces, hydrogen gas should be used as an alternative reducing agent in the ironmaking process in the presence of argon gas. Moreover, the use of hydrogen naturally eliminated sulfur and also speed up the reduction rate of the base material. Carbothermal reduction of iron oxide is less pure because of the high chemical activity of carbon monoxide and its inability to exist longer in its free state. CO's ability to exist as a transition metal entails it can also exist in carbide forms. Thus, to mitigate the aftermath effect of challenges caused by the use of $CO/CO_2$ on iron oxide during reduction, the use of hydrogen [77], water gas [78], and hydrocarbon [79] have been encouraged in recent works of literature [18, 19, 47, 80]. Therefore, hydrogen usage in the iron and steelmaking industry remains a prominent way to curtail the menace of the greenhouse effect, energy loss, and global warming in our immediate environment.

### 4.2 Reduction by methane gas and other hydrocarbons

The reduction of iron ore by carbon monoxide can produce carbondioxide which is utilized by the ecosystems for anaerobic respiration and photosynthetic operations [80]. Consequentially, the oxidation of methane gas and other hydrocarbons does produce water gas (CO and $H_2$) which is charged into a fluidized bed reactor for iron and steel production. Nasr et al. [71] used the chemical looping combustion method to investigate the reduction kinetics of iron ore in a methane atmosphere. The increase in $CO_2$ concentration in the atmosphere has attracted considerable interest in the metallurgical industry. Thus, the use of methane gas as an alternative source of carbon-bearing material has been uninitialized as a promising concept for capturing $CO_2$ in the atmosphere and for power generation via the chemical looping combustion (CLC) method. The reduction of iron ore in the methane atmosphere may not follow the usual stepwise order (i.e. it is either the magnetite stage or the wustite state is being bypassed). This implies the rate of reaction will be faster in the iron oxide/methane reduction reaction as the burning of methane gas in the absence of air produces a chemical loop of CO and $H_2$ which makes the reduction reaction of iron oxide to iron proceed much faster into completion compared to when using the individual reducing gas (i.e. CO and $H_2$). Thus, the reduction of hematite ($\alpha$-$Fe_2O_3$) and maghemite ($\gamma$-$Fe_2O_3$) in the $CH_4$ atmosphere at standard temperature and pressure in the absence of air will produce iron in the following chronological order $Fe_2O_3 \rightarrow Fe_3O_4 \rightarrow Fe$ and $Fe_2O_3 \rightarrow FeO \rightarrow Fe$ respectively [35, 66]. This reduction method provides good support performance, increases mechanical strength, and provides attrition resistance for metallic iron produced under environmentally safe conditions and economically viable. Thus, for increased productivity and a high reduction rate, the direct reduction of iron ore with methane does have intense catalytic nature that can be used in the powder metallurgical process.

### 4.3 Reduction in biomass gas

Biomass is a common solid fuel, and its pyrolysis products, with highly reductive characteristics which occur in solid, liquid, and gaseous states [81]. The quest for coal substitutes in the iron ore reduction process has attracted considerable interest, as the use of pyrolysis products can be viewed as an alternative reductant. This need has its implication to acquire iron ore reduction by metallurgical industries which possess good quality carbon-bearing properties with reduction characteristics, easy renewability, lower greenhouse effect, and fewer energy-saving challenges. The research on pine saw dust usage, as a reductant for pure $Fe_2O_3$ particles was investigated by Liu et al. [72]. Biomass pyrolysis of pure $Fe_2O_3$ was conducted on pine sawdust at a mass ratio of 3:2 at a calcination time of 15mins. The mixture was fired at a reduction temperature range of 350 – 550°C under a nitrogen-regulated atmosphere. Iron oxide reduced with biomass sawdust can produce pollution-free pure metallic iron. Few research works have attempted to study the use of biomass in the ironmaking process. Gan et al. [82] examined the pollutant effect on sintered iron ore reduced using biomass fuel as a reductant. Luo et al. [83] uses biomass syngas to perform direct reduction on hematite-biomas briquette. The briquette powdered ore-biomass was heated at a reduction temperature range of 450 to 900°C. Strezov, et al. [84, 85], performed iron ore reduction runs using biomass sawdust. The iron was reduced from 30% sawdust to predominantly metallic iron at a reduction temperature above 600°C. The reduction of the iron ore-sawdust mixture begins precisely at 670°C and ended at 1200°C. The implication of using biomass in the ironmaking process is that the process is less energy-consuming, sulfur free, and environmentally friendly without causing the greenhouse effect.

Furthermore, due to the vast utilization capacity of the different iron processing methods in modern-day iron and steel-making technology, it becomes imperative and advantageous to enumerate the basic challenges and limitations peculiar to convectional iron and steel production routes by proferring meaningful solutions and mapping-out holistic future perspectives to issue surrounding extraction process of pure metallic iron. Thus, this study suggests the need to employ a more pragmatic metallurgical extraction approach in the production routes for iron processing with lower production costs and minimal energy requirements [86, 87]. This can also be attributed to the ability to utilize other alternative energy sources instead of the use of metallurgical coke as a reductant. This will lower the primary operation cost, energy consumption, and elimination of operational tendencies for metallurgical coke, and pre-reduction processes (i.e. beneficiation, pelletization, recycling, etc.) may no longer be necessary, the SR process utilizes direct fines and sinters as feeds, better energy conservative process with rich off gas used for power generation, better control of production process parameters (temperature, gas flow rate, reduction rate, etc.). Faster reduction rate and quick iron recovery are certain, as good thermo-chemical configuration and flexibility in hot metal-slag separation, faster reaction kinetics due to high smelting intensity and productivity, etc. [13, 88]. In contrast, the SR process also has some limitations despite its economic viability and high productivity tendencies. This includes high oxygen and power consumption requiring mandatory ore fine pre-reduction process and the need to meet up with required energy for a highly efficient post-combustion process.

## 5. Future research perspective

In the quest to improve the quality of iron produced across the globe, so many works have been done through the development of various ironmaking and steel-making techniques. Many ideas have been developed on how to improve the various metallurgical operations governing the extraction process of iron and steel research all over the world. These iron extraction processes are mainly categorized into the blast furnace, direct reduction, indirect reduction, and reduction-smelting processes. Metallurgical factors such as reduction agent types, quality of iron oxide, energy sources and analysis, cost of production, etc. are scarcely considered in the works of literature. Thus, it is expedient that future research perspectives should consider the aforementioned factors affecting the overall metallurgical extraction process in iron and steel production before active application in extraction technique. In addition, the influence of foreign bodies in the metal matrix of iron metals produced by the different extraction processes should be checked and excess silicate and aluminate with the banded microstructure of iron oxide metal matrix will require the use of the excessive fluxing agent, increased energy demand during the iron smelting process. Thus, further studies on the effect of gangue elements on the iron metal quality in convectional iron and steelmaking processes should be further investigated.

## 6. Conclusion

The fast depletion of metallurgical coke in the present era iron and steelmaking making process remains an eye-opener for metallurgical experts across the globe. This challenge has reiterated the need to seek other alternative energy sources for the iron and steelmaking process with viable production routes. Due to economic implications, complex metallurgical operations, high energy demand, and time-consuming tendencies that accompany the conventional blast furnace/blast oxygen furnace operation and direct reduction route. This study provides an extensive overview of selected iron and steelmaking processes with special emphasis on the various alternative energy sources that are used instead of conventional metallurgical coke. Thus, the study concludes as follows:

1. Apart from the traditional method of iron production using the blast furnace process, the direct reduction process, indirect reduction process, and smelting-reduction process remains the most recent trend used in present-era ironmaking technology.

2. The use of gaseous reductants (i.e. CO, $H_2$, $CH_4$, etc.) remains a viable alternative energy source provided they can be acquired at a cheaper rate as the usage can be unsustainable when applied in iron and steelmaking processes at an industrial scale.

3. The operational capabilities of the conventional direct smelting process can be improved upon when modified with a post-reduction process within its ambiance provided the use of energy saving approach is encouraged.

4. The non-contact direct reduction technique has shown its viability for decarburization of direct reduct iron with minimal energy requirement and it has shown to be cost-effective.

5. The smelting-reduction technique has an advantage over direct reduction and indirect reduction processes as it holds the capacity to utilize many available energy sources which makes it more time-saving, less energy-consuming, and operationally efficient.

6. The use of gaseous reductants in the smelting-reduction process of iron oxide remains a more promising approach in the production of pure molten iron, free from unwanted impurities within the phase microstructure of the by-product.

## Acknowledgements

This work was supported by the Petroleum Technology Development Fund, Nigeria. Grant Number: PTDF/ED/LSS/PhD/JEO/0278/19, 2019.

## Conflict of interest

The authors declare no conflict of interest.

**Author details**

Joseph Ekhebume Ogbezode[1,2]*, Olusegun Olufemi Ajide[1],
Oluleke Olugbemiga Oluwole[1] and Olusoji Ofi[1]

1 Department of Mechanical Engineering, University of Ibadan, Ibadan, Nigeria

2 Department of Mechanical Engineering, Edo State University, Uzairue, Edo State, Nigeria

*Address all correspondence to: ogbezode.joseph@edouniversity.edu.ng

**IntechOpen**

© 2023 The Author(s). Licensee IntechOpen. This chapter is distributed under the terms of the Creative Commons Attribution License (http://creativecommons.org/licenses/by/3.0), which permits unrestricted use, distribution, and reproduction in any medium, provided the original work is properly cited.

## References

[1] Ahmed H. New trends in the application of carbon-bearing materials in blast furnace iron-making. Minerals. 2018;**8**(12):561. DOI: 10.3390/min8120561

[2] Abdus-Salam N, Ugbe FA, Funtua MA. Characterization of synthesized goethite and natural goethite sourced from Itakpe in north central, Nigeria. Chemistry Search Journal. 2018, 2018;**9**(2):24-32

[3] Abdel Halim KS. Isothermal reduction behavior of $Fe_2O_3$/MnO composite materials with solid carbon. Materials Science and Engineering: A. 2007;**452-453**:15-22. DOI: 10.1016/j.msea.2006.12.126

[4] Alamsari B, Torii S, Trianto A, Bindar Y. Heat and mass transfer in reduction zone of sponge iron reactor. Journal International Scholarly Research Network, ISRN Mechanical Engineering. 2011:1-12. DOI: 10.5402/2011/324659

[5] Siddiqi H, Chandaliya VK, Suresh A, Dash PS, Meikap BC. A scale-up approach to produce highly reactive iron ore catalyzed coke for blast furnace operation. Mineral Processing and Extractive Metallurgy Review. 2019;**41**:1-11. DOI: 10.1080/08827508.2019.1666124

[6] Babich A, Mousa EA, Senk D. Reduction behavior of iron ore pellets with simulated coke oven gas and natural gas. Steel Research International. 2013;**84**(11):1085-1097. DOI: 10.1002/srin.201200333

[7] Bahgat M, Abdel Halim KS, El-Kelesh HA, Nasr MI. Blast furnace operating conditions manipulation for reducing coke consumption and $CO_2$ emission. Steel Research International. 2012;**83**(7):686-694. DOI: 10.1002/srin.201200001

[8] Angalakuditi VB, Bhadravathi P, Gujare R, Ayyappan G, Singh LR. Mineralogical aspects of reducing lump iron ore, pellets, and sinter with hydrogen. Metallurgical and Materials Transactions B. 2022;**53**:1036-1065. DOI: 10.1007/s11663-022-02442-4

[9] Arisanti R, Yusuf M, Faizal M. Study of the effect of proximate, ultimate, and calorific value analysis on methane gas emission (CH4) on combustion of coal for sustainable environment. Science and Technology Indonesia. 2018;**3**(2):100-106. DOI: 10.26554/sti.2018.3.2.100-106

[10] Romenets VA, Usachev AB, Balasanov AV, Lekherzak VE. The role of coal in the Romelt process for the liquid-phase reduction of iron. Metallurgist. 2001;**45**(3/4):101-105. DOI: 10.1023/a:1010555923457

[11] Roy S, Das A. Characterization and processing of low-grade iron ore slime from the Jilling area of India. Mineral Processing and Extractive Metallurgy Review. 2008;**29**(3):213-231. DOI: 10.1080/08827500801997886

[12] Sah R, Dutta SK. Smelting reduction of iron ore-coal composite pellets. Steel Research International. 2010;**81**(6):426-433. DOI: 10.1002/srin.200900113

[13] Shekhawat D, Kukshal V, Banerjee MK, Patnaik A. Study the reduction of mill scale with lean grade coal through RI-RDI. IOP Conference Series: Materials Science and Engineering. 2021;**1017**:1-13. DOI: 10.1088/1757-899X/1017/1/012037

[14] Sharma MK, Solanki V, Roy GG, Sen PK. Study of reduction behaviour of prefabricated iron ore–graphite/coal composite pellets in rotary hearth furnace. Journal of Ironmaking and Steelmaking. 2013;**40**:590-597. DOI: 10.1179/1743281212Y.0000000086

[15] Wan X, Shen L, Jokilaakso A, Eriç H, Taskinen P. Experimental approach to matte–slag reactions in the flash smelting process. Mineral Processing and Extractive Metallurgy Review. 2020;**42**: 1-11. DOI: 10.1080/08827508.2020.1737801

[16] Wang H, Sohn HY. Effects of firing and reduction conditions on swelling and iron whisker formation during the reduction of iron oxide compact. ISIJ International. 2011;**51**(6):906-912. DOI: 10.2355/isijinternational.51.906

[17] Wang HT, Sohn HY. Effect of CaO and SiO2 on swelling and iron whisker formation during reduction of iron oxide compact. Ironmaking & Steelmaking. 2011;**38**(6):447-452. DOI: 10.1179/1743281211y.0000000022

[18] Mousa EA, Babich A, Senk D. Utilization of coke oven gas and converter gas in the Direct Reduction of lump iron ore. Metallurgical and Materials Transactions B. 2013;**45**(2): 617-628. DOI: 10.1007/s11663-013-9978-6

[19] Qie J, Zhang Y, Wang J, Zhang X, Fu T, Cui K. Reduction and melting behavior of pre-reduction metallized carbon-bearing pellets in iron Bath smelting reduction. Journal of Materials. 2022;**74**:1-10. DOI: 10.1007/s11837-021-05104-z

[20] Liu Z, Bi X, Gao Z, Liu W. Carbothermal reduction of iron ore in its concentrate-agricultural waste pellets. Advances in Materials Science and Engineering. 2018;**2018**:1-6. DOI: 10.1155/2018/2138268

[21] El-Geassy AA, Nasr MI, El-Raghy SM, Hammam AA. Comparative studies on isothermal and non-isothermal reduction of Haematite in carbon monoxide atmosphere. Ironmaking & Steelmaking. 2019;**47**:1-10. DOI: 10.1080/03019233.2019.1646564

[22] Donskoi E, McElwain DL, Wibberley LJ. Estimation and modelling of parameters for direct reduction in iron ore/coal composites: Part II. Kinetic parameters. Journal of Metallurgy and Materials Transaction B. 2003;**34**:255-266. DOI: 10.1007/s11663-003-0012-2

[23] Donskoi E, Olivares RI, McElwain DL, Wibberley LJ. Experimental study of coal based direct reduction in iron/coal composite pellets in a one layer bed under non-isothermal asymmetric heating. Journal of Ironmaking and Steelmaking. 2006;**33**: 24-28. DOI: 10.1179/174328106X80064

[24] Hasanbeigi A, Arens M, Price L. Alternative emerging ironmaking technologies for energy efficiency and carbon dioxide emissions reduction: A technical review. Renewable and Sustainable Energy Reviews. 2014;**33**: 645-658. DOI: 10.1016/j.rser.2014.02.031

[25] Ramakgala C, Danha G. A review of ironmaking by direct reduction processes: Quality requirements and sustainability. Procedia Manufacturing. 2019;**35**:242-245. DOI: 10.1016/j.promfg.2019.05.034

[26] Folorunso DO, Olubambi P, Borode JO. Characterization and qualitative analysis of some Nigerian clay deposits for Rafractory applications. IOSR Journal of Applied Chemistry (IOSR-JAC). 2014;7(9):40-47. DOI: 10.9790/5736-7914047

[27] Anameric B, Kawatra SK. Direct iron smelting reduction processes, mineral processing, and extractive metallurgy review: An. International Journal. 2008; **30**(1):1-51. DOI: 10.1080/0882750080 2043490

[28] Gojic M, KoÞuh S. Development of direct reduction processes and smelting reduction processes for the steel production. Journal of Chemists and Chemical Engineers. 2022;**55**(1):1-10

[29] Ogbezode JE, Oluwole OO. Reduction kinetics behavior of goethite iron ore in the CO/CO2 atmosphere from wood charcoal. International Journal of Emerging Engineering Research and Technology (IJEERT). 2019;7(4):11-21

[30] Ogbezode J, Ajide O, Ofi O, Oluwole OO. Determination of the reaction rate controlling resistance of goethite iron ore reduction using CO/ $CO_2$ gases from wood charcoal. Material Proceedings, MPDI. 2021;**3**(1):27. DOI: 10.3390/IEC2M-09373

[31] Ogbezode J, Ajide O, Ofi O, Oluwole O. An overview of the reduction-smelting process of iron oxides in modern-day ironmaking technology. Research and Development in Material Science (RDMS). 2022;**17**(4): 1992-1994. DOI: 10.31031/ RDMS.2022.17.000918

[32] Muwanguzi AJB, Karasev AV, Byaruhanga JK, Jönsson PG. Characterization of chemical composition and microstructure of natural iron ore from Muko deposits. ISRN Materials Science. 2012:1-9. DOI: 10.5402/2012/174803

[33] Zhu D, Luo Y, Pan J, Zhou X. Characterization of semi-coke generated by coal-based direct reduction process of siderite. Journal of Central South University. 2015;**22**(8):2914-2921. DOI: 10.1007/s11771-015-2826-x

[34] Huitu K, Helle M, Helle H, Kekkonen M, Saxén H. Optimization of Midrex direct reduced iron use in ore-based steelmaking. Steel Research International. 2014;**86**(5):456-465. DOI: 10.1002/srin.201400091

[35] Heidari A, Niknahad N, Iljana M, Fabritius TA. Review on the kinetics of iron ore reduction by hydrogen. Materials. 2021;**14**:7540. DOI: 10.3390/ ma14247540

[36] Li K, Wen N, Zhu M, Zheng M, Yuan L. Iron extraction from Oolitic iron ore by a deep reduction process. Journal of Iron and Steel Research International. 2011;**18**(8):9-13. DOI: 10.1016/ s1006-706x(11)60096-4

[37] Bhaskar A, Assadi M, Nikpey SH. Decarbonization of the iron and steel industry with direct reduction of iron ore with green hydrogen. Energies. 2020;**13**(3):758. DOI: 10.3390/ en13030758

[38] Gonoring TB, Franco AR Jr, Vieira EA, Nascimento RC. Kinetic analysis of the reduction of hematite fines by cold hydrogen plasma. Journal of Materials Research and Technology. 2022;**20**:2173-2187. DOI: 10.1016/j. jmrt.2022.07.174

[39] Hou B, Zhang H, Li H, Zhu Q. Study on kinetics of iron oxide reduction by hydrogen. Chinese Journal of Chemical Engineering. 2012;**20**(1):10-17. DOI: 10.1016/s1004-9541(12)60357-7

[40] Du W, Yang S, Pan F, Shangguan J, Lu J, Liu S, et al. Hydrogen reduction of hematite ore fines to magnetite ore fines at low temperatures. Journal of Chemistry. 2017:1-11. DOI: 10.1155/2017/ 1919720

[41] Vogl V, Åhman M, Nilsson LJ. Assessment of hydrogen direct reduction for fossil-free steelmaking. Journal of Cleaner Production. 2018;**203**:736-745. DOI: 10.1016/j.jclepro.2018.08.279

[42] Ma Y, Souza FIR, Zhang X, Nandy S, Barriobero-Villa P, Requena G, et al. Hydrogen-based direct reduction of iron oxide at 700°C: Heterogeneity at pellet and microstructure scales. International Journal of Minerals, Metallurgy, and Materials. 2022;**29**(10):1901-1907. DOI: 10.1007/s12613-022-2440-5

[43] Naseri Seftejani M, Schenk J, Zarl MA. Reduction of Haematite using hydrogen thermal plasma. Materials. 2019;**12**(10):1608. DOI: 10.3390/ma12101608

[44] Naseri Seftejani M, Schenk J. Thermodynamics of liquid iron ore reduction by hydrogen thermal plasma. Metals. 2018;**8**(12):1051. DOI: 10.3390/met8121051

[45] Naseri Seftejani M, Schenk J, Spreitzer D, Andreas ZM. Slag formation during reduction of iron oxide using hydrogen plasma smelting reduction. Materials. 2020;**13**(4):935. DOI: 10.3390/ma13040935

[46] Fowkes N. The reduction of iron ore pellets used in the FASTMET process. IMA Journal of Management Mathematics. 1999; **10**(1):27-39. DOI: 10.1093/imaman/10.1.27

[47] Youling L, Huibin L, Hua W, Shan Q, Jianhang H, Yali H, et al. Smelting potential of HIsmelt Technology for High-Phosphorus Iron ore and Ilmenite. In: 2011 International Conference on Computer Distributed Control and Intelligent Environmental Monitoring. IEEE Computer Society. 2011. DOI: 10.1109/cdciem.2011.353

[48] Shi B, Zhu D, Pan J, Hu B, Wang Z. Reducing process of sinter in COREX shaft furnace and influence of sinter proportion on reduction properties of composite burden. Journal of Central South University. 2021;**28**(3):690-698. DOI: 10.1007/s11771-021-4638-5

[49] Agrawal S, Dhawan N. Microwave Carbothermic reduction of low-grade iron ore. Metallurgical and Materials Transactions B. 2020;**51**:1576-1586. DOI: 10.1007/s11663-020-01883-z

[50] Murakami T, Shinomiya H, Maruoka D, Kasai E. Effect of types of carbonaceous material and CaO addition on reduction behavior of pre-reduced iron ore–carbon composite. ISIJ International. 2019;**59**(6):1011-1017. DOI: 10.2355/isijinternational.isijint-2018-725

[51] Khasraw D, Spooner S, Hage H, Meijer K, Li Z. Devolatilisation characteristics of coal and biomass with respect to temperature and heating rate for HIsarna alternative ironmaking process. Fuel. 2021;**284**:119101. DOI: 10.1016/j.fuel.2020.119101

[52] Zhang L, Zhu Y, Yin W, Guo B, Rao F, Ku J. Isothermal coal-based reduction kinetics of Fayalite in copper slag. ACS Omega. 2020;**5**(15):8605-8612. DOI: 10.1021/acsomega.9b04497

[53] Li H, Pinson DJ, Zulli P, Lu L, Longbottom RJ, Chew SJ, et al. Interaction between mineral phases in a hematite iron ore and fluxing materials during sintering. Metallurgical and Materials Transactions B. 2020;**52**: 267-281. DOI: 10.1007/s11663-020-02010-8

[54] Kumar M, Baghel H, Patel KS. Reduction and swelling of fired hematite iron ore pellets by non-coking coal fines for application in sponge ironmaking.

Mineral Processing and Extractive Metallurgy Review: An International Journal. 2013;**34**(4):249-267. DOI: 10.1080/08827508.2012.656776

[55] Kumar M, Patel SK. Characteristics of Indian non-coking coals and iron ore reduction by their chars for directly reduced iron production. Mineral Processing and Extractive Metallurgy Review. 2008;**29**(3):258-273. DOI: 10.1080/08827500801997902

[56] Pal J, Arunkumar C, Rajshekhar Y, Das G, Goswami MC, Venugopalan T. Development on iron ore Pelletization using calcined lime and MgO combined flux replacing limestone and bentonite. ISIJ International. 2014;**54**(10):2169-2178. DOI: 10.2355/isijinternational.54.2169

[57] Haque R, Ray HS. Role of ore/carbon contact and a direct reduction in the reduction of iron oxide by carbon. Metallurgical and Materials Transactions B. 1995;**26**(2):400-401. DOI: 10.1007/bf02660982

[58] Fortini OM, Fruehan RJ. Rate of reduction of ore-carbon composites: Part I. determination of intrinsic rate constants. Metallurgical and Materials Transactions B. 2005;**36**(6):865-872. DOI: 10.1007/s11663-005-0088-y

[59] Onyedika G, Nwoko C, Oguarah A, Ogwuegbu M. Comparative kinetics of iron ore dissolution in aqueous HCl-$HNO_3$ system. Journal of Minerals and Materials Characterization and Engineering. 2013;**1**:153-159. DOI: 10.4236/jmmce.2013.14026

[60] Odokuma-Alonge O. A geochemical appraisal of some marble physiques in Ubo area and environs, southwestern Nigeria. Journal of Applied Sciences and Environmental Management. 2019;**23**(6):1189. DOI: 10.4314/jasem.v23i6.29

[61] Jimoh OA, Ariffin KS, Hussin HB, Habeeb AA. Characterization and assessment of Okpella Metacarbonate deposit in Nigeria. Carbonates and Evaporites. 2016;**32**(4):513-524. DOI: 10.1007/s13146-016-0308-3

[62] Oluwatoyin OA, Olusola AO. Lithological features and chemical characterization of metamorphosed carbonate rocks in Igue, southwestern, Nigeria. Journal of Geology and Mining Research. 2021;**13**(1):11-20. DOI: 10.5897/JGMR2020.0349

[63] Olayebi OO. Steelmaking experience in the use of Nigerian in ore at the Delta steel company, Nigeria. Journal of Chemical Engineering and Materials Science. 2014;**5**(5):47-62. DOI: 10.5897/JCEMS2014.0166

[64] Omang BO, Kudamnya EA, Owolabi AO, Odey J, Aniwetalu EU, Ako TA. Characterization of kaolin deposits in Okpella and environs southern Nigeria. International Journal of Geosciences. 2019;**10**:317-327. DOI: 10.4236/ijg.2019.103018

[65] Omole SO, Oyetunji A, Alaneme KK, Olubambi PA. Structural characterization and mechanical properties of pearlite-enhanced micro-alloyed ductile irons. Journal of King University-Engineering Sciences. 2018;**32**(3):205-210. DOI: 10.1016/j.jksues.2018.11.008

[66] Lu C, Li K, Wang H, Zhu X, Wei Y, Zheng M, et al. Chemical looping reforming of methane using magnetite as oxygen carrier: Structure evolution and reduction kinetics. Applied Energy. 2018;**211**:1-14. DOI: 10.1016/j.apenergy.2017.11.04

[67] Subhnit KR, Deepak NND, Swagat SR. Influence of coal petrography on microwave-assisted Carbothermic reduction roasting of banded hematite Jasper ore. Mineral Processing and Extractive Metallurgy Review. 2020;**42, 2020**:1-15. DOI: 10.1080/08827508.2020.1743290

[68] Liu G, Strezov V, Lucas JA, Wibberley LJ. Thermal investigations of direct iron ore reduction with coal. Thermochimica Acta. 2004;**410**(1–2):133-140. DOI: 10.1016/s0040-6031(03)00398-8

[69] Hammam A, Cao Y, El-Geassy HA, El-Sadek MH, Li Y, Wei H, et al. Non-isothermal reduction kinetics of iron ore fines with carbon-bearing materials. Metals. 2021;**11**:1137. DOI: 10.3390/met11071137

[70] Li S, Zhang H, Nie J, Dewil R, Baeyens J, Deng Y. The direct reduction of iron ore with hydrogen. Sustainability. 2021;**13**:8866. DOI: 10.3390/su13168866

[71] Nasr S, Plucknett KP. Kinetics of iron ore reduction by methane for chemical looping combustion. Energy & Fuels. 2014;**28**(2):1387-1395. DOI: 10.1021/ef402142q

[72] Liu X, Zhang H, Li S, Li D, Huang D. Research on reduction of $Fe_2O_3$ by biomass sawdust. Journal of Shanghai Jiaotong University (Science). 2017;**22**(3):280-285. DOI: 10.1007/s12204-017-1833-5

[73] Speight GJ. Ultimate analysis. In: Handbook of Coal Analysis. John Wiley & Sons; 2015. pp. 144-169. DOI: 10.1002/9781119037699.ch6

[74] Htet TT, Yan Z, Spooner S, Degirmenci V, Meijer K, Li Z. Gasification and physical-chemical characteristics of carbonaceous materials in relation to HIsarna ironmaking process. Fuel. 2021;**289**:119890. DOI: 10.1016/j.fuel.2020.119890

[75] Bielowicz B. Petrographic characteristics of coal gasification and combustion by-products from high volatile bituminous coal. Energies. 2020;**13**(17):4374. DOI: 10.3390/en13174374

[76] Xing L, Qu Y, Wang C, Shao L, Zou Z. Gas–liquid reduction behavior of hematite ore fines in a flash reduction process. Metallurgical and Materials Transactions B. 2022;**51**:1233-1242. DOI: 10.1007/s11663-020-01811-1

[77] Pineau A, Kanari N, Gaballah I. Kinetics of reduction of iron oxides by $H_2$. Thermochimica Acta. 2006;**447**(1):89-100. DOI: 10.1016/j.tca.2005.10.004

[78] Pavlov AI. Coal gasification as an alternative to natural gas. Steel in Translation. 2009;**39**(6):475-478. DOI: 10.3103/s0967091209060096

[79] Palacios P, Toledo M, Cabrera M. 2015. Iron ore reduction by methane partial oxidation in a porous media. International Journal of Hydrogen Energy. 2015;**40**(31):9621-9633. DOI: 10.1016/j.ijhydene.2015.05.058

[80] Pavlína P, Simona J. Process engineering in iron production, journal of. Chemical and Process Engineering. 2013;**34**:63-76. DOI: 10.2478/cpe-2013-0006

[81] Brázda P, Kohout J, Bezdička P, Kmječ T. $\alpha$-$Fe_2O_3$ versus $\beta$-$Fe_2O_3$: Controlling the phase of the transformation product of $\varepsilon$-$Fe_2O_3$ in the $Fe_2O_3/SiO_2$ system. Crystal Growth & Design. 2014;**14**(3):1039-1046. DOI: 10.1021/cg4015114

[82] Gan M, Fan X, Chen X, Ji Z, Lv W, Wang Y, et al. Reduction of pollutant

emission in iron ore sintering process by applying biomass fuels. ISIJ International. 2012;**52**(9):1574-1578. DOI: 10.2355/isijinternational.52.1574

[83] Luo S, Yi C, Zhou Y. Direct reduction of mixed biomass-Fe2O3 briquettes using biomass-generated syngas. Renewable Energy. 2011;**36**(12): 3332-3336. DOI: 10.1016/j.renene.2011.05.006

[84] Strezov V. Iron ore reduction using sawdust: Experimental analysis and kinetic modeling. Renewable Energy. 2006;**31**(12):1892-1905. DOI: 10.1016/j.renene.2005.08.032

[85] Tleugabulov SM, Abikov SB, Altybaeva DK, Isupov YD, Tleugabulov BS. Reductive smelting of iron ore. Steel in Translation. 2015; **45**(5):351-355. DOI: 10.3103/s0967091215050162

[86] Tleugabulov SM, Eremeeva ZV, Kurmanseitov MB. Carburization and decarburization in iron and steel production. Steel in Translation. 2018; **48**(11):732-736. DOI: 10.3103/s0967091218110116

[87] Zhang Y, Gao Q, Zhao J, Li M, Qi Y. Semi-smelting reduction and magnetic separation for the recovery of iron and alumina slag from iron rich bauxite. Minerals. 2019;**9**(4):223. DOI: 10.3390/min9040223

[88] Ofoegbu SU. Characterization studies on Agbaja iron ore: A high-phosphorus content ore, SN. Applied Sciences. 2019;**1**:204. DOI: 10.1007/s42452-019-0218-9